Functional Coordination Polymers and Metal–Organic Frameworks

Functional Coordination Polymers and Metal–Organic Frameworks

Editor

Andrea Rossin

MDPI • Basel • Beijing • Wuhan • Barcelona • Belgrade • Manchester • Tokyo • Cluj • Tianjin

Editor
Andrea Rossin
Istituto di Chimica dei Composti
Organometallici (ICCOM)
Consiglio Nazionale delle
Ricerche (CNR)
Sesto Fiorentino (Firenze)
Italy

Editorial Office
MDPI
St. Alban-Anlage 66
4052 Basel, Switzerland

This is a reprint of articles from the Special Issue published online in the open access journal *Inorganics* (ISSN 2304-6740) (available at: www.mdpi.com/journal/inorganics/special_issues/CPs_MOFs).

For citation purposes, cite each article independently as indicated on the article page online and as indicated below:

LastName, A.A.; LastName, B.B.; LastName, C.C. Article Title. *Journal Name* **Year**, *Volume Number*, Page Range.

ISBN 978-3-0365-1500-7 (Hbk)
ISBN 978-3-0365-1499-4 (PDF)

Contents

About the Editor

Andrea Rossin

Andrea Rossin [*h*-index (ISI-WOS) = 30] was born in Biella, Northern Italy, in 1974. In 2001-2004 he developed his Ph.D. work at Cardiff University (Wales, UK), under the supervision of Prof. Simon Aldridge. In 2005 he was a post-doctoral fellow at the Universitat Autònoma de Barcelona (Barcelona, Spain) in the group of Prof. Agustì Lledòs. In 2006, he moved to ICCOM-CNR. Since January 2021 he is a Senior Researcher in the same institute. His research interests are twofold: (a) formate-, thiazole-, and thiazolidine-based Metal Organic Frameworks (MOFs) synthesis, characterization and applications in gas storage, catalysis and luminescence sensing; (b) reactivity studies of ammonia-borane and amino-boranes with transition metal organometallics (chemical hydrogen storage). He has co-authored ca. 110 papers published on international peer-reviewed journals.

Preface to "Functional Coordination Polymers and Metal–Organic Frameworks"

When I was asked to prepare a Special Issue of *Inorganics* focused on Coordination Polymers and Metal-Organic Frameworks, I could not imagine that it would have such a great success and participation. I am really pleased to have gathered so many excellent contributions coming from three different continents all over the World: America, Europe, and Asia. The topics covered are diverse, including synthesis, advanced characterization, and applications in the fields of CO_2 storage, water adsorption, heterogeneous catalysis, and luminescence. The incredible flexibility in CPs and MOFs design stemming from a virtually infinite combination of organic linkers of assorted nature (azolates, carboxylates) and metallic nodes containing transition metals with specific chemical properties opens wide horizons in the future development of these exciting materials. I hope that the *Inorganics* readers will enjoy reading this Special Issue as much as I did when I assembled it in its final stage. I must acknowledge once more all the contributors (some of them are also friends!) and the *Inorganics* Editorial Office for its valuable support in this happy-ending adventure.

Andrea Rossin
Editor

inorganics

MDPI

Editorial

Editorial for Special Issue "Functional Coordination Polymers and Metal–Organic Frameworks"

Andrea Rossin

Istituto di Chimica dei Composti Organometallici (ICCOM-CNR), Via Madonna del Piano 10, 50019 Sesto Fiorentino, Italy; arossin@iccom.cnr.it

check for updates

Citation: Rossin, A. Editorial for Special Issue "Functional Coordination Polymers and Metal–Organic Frameworks". *Inorganics* **2021**, *9*, 33. https://doi.org/10.3390/inorganics9050033

Received: 12 April 2021
Accepted: 28 April 2021
Published: 3 May 2021

Metal–Organic Frameworks (MOFs) and Coordination Polymers (CPs) are at the forefront of contemporary coordination chemistry research, as witnessed by the impressive (and ever-growing) number of publications appearing in the literature on this topic in the last 20 years (Figure 1), reaching almost 4000 papers in 2020. Among them, several recent review articles and books clearly illustrate the huge potential of this class of compounds [1–6].

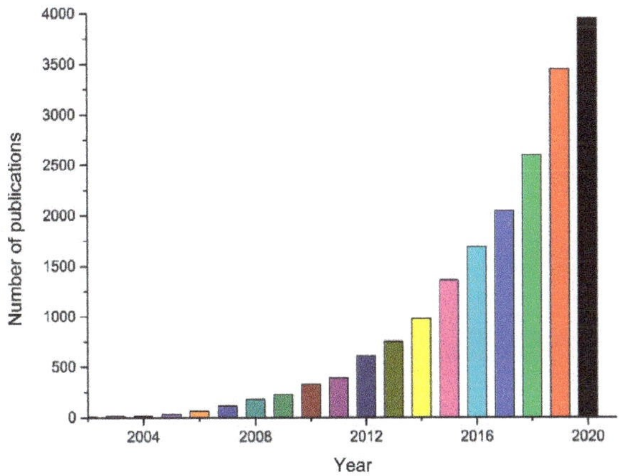

Figure 1. The number of publications in the 2002–2020 time period whose titles or keywords contain "MOF". Data from Web of Science.

The virtually infinite combination of tailored organic linkers and inorganic metal nodes (in the form of isolated ions or more complex cluster structures) generates a wide variety of functional materials for assorted applications. With a specific application in mind, chemists and materials scientists devoted to MOFs synthesis "play LEGO©" [7] every day in their laboratories to reach their targets, choosing the most suitable pairs to create valuable MOFs.

The present Special Issue collects a total number of 10 papers: 8 original articles and 2 review articles covering multiple aspects of MOFs and CPs chemistry.

Two fundamental synthetic studies are part of this collection. Armelao et al. (Padova, Italy) [8] examined the solvent effects in the structures of Cu^{II}-4,4′-bipyridyl CPs coupled to different coordinating solvents, such as DMA, DMF and DMSO. The CP dimensionality (either 2D sheet or 1D chain) strongly depends on the solvent of choice. Thus, the coordinating solvent approach (CSA) can be used as an effective tool to modulate and control the dimensionality, composition and network of CPs. Galli and co-workers (Como,

Italy) [9] proved the versatility of the 1,3-bis(1,2,4-triazol-4-yl)adamantane (tr_2ad) linker in the synthesis of new Cd^{II}, Cu^{II} and Zn^{II} CPs of assorted dimensionality (ranging from 1D to 2D) and featured by excellent thermal stability. The high thermal robustness is quite common in MOFs and CPs built with N-donors like pyrazolates, triazolates and imidazolates, and it stems from the strength of the metal-nitrogen coordination bond.

This Special Issue though is focused on applications, collecting contributions covering many different applicative contexts. In the field of heterogeneous catalysis, an excellent and comprehensive review article by Zhou et al. (College Station, TX, USA) [10] summarizes the most recent advances in organometallic functionalization of MOFs from both linker-centric and metal-cluster-centric perspectives. MOFs can function as a tailorable platform for traditional organometallic transformations, including reaction of alkenes, cross-coupling reactions and C–H activations.

The timely CO_2 adsorption topic [at the core of contemporary Carbon Capture and Sequestration (CCS) technology to reduce carbon dioxide levels in the Earth's atmosphere and concomitant greenhouse effect] is developed by Bonino and co-workers (Torino, Italy) [11]. The research team in Torino reports on the synthesis, spectroscopic characterization and adsorption properties of the Ce^{III} MOF $Ce_5(BDC)_{7.5}(DMF)_4$, containing the simple terephtalic acid (H_2BDC) as linker. The coordinated DMF solvent can be removed from the metal nodes generating open metal sites, as confirmed by FTIR spectroscopy. The interaction of the desolvated material with CO_2 was characterized by volumetric and calorimetric measurements, finding a relevant heat of adsorption (Q_{st}) for the very first dose, typical of MOFs with open metal sites. Another contribution focused on the same topic comes from Taddei et al. (Swansea, Wales, UK) [12]. The team reported a systematic approach aimed at identifying the optimal conditions for the compaction (i.e., pellets preparation) of MOF-801, a small-pore zirconium-based MOF containing fumaric acid as linker. The MOF pellets (prepared under different experimental pressure conditions and in the presence of assorted binders) were tested in CO_2 adsorption, retaining as much as 90% of the CO_2 working capacity of the powder, while displaying unaffected sorption kinetics. This work provides a starting point for future exploration of shaping of MOF powders, which is becoming a progressively more important aspect as MOFs move towards commercialization and employment in industrial applications.

Another application of growing interest for MOFs and CPs is luminescence. As chemical sensors, luminescent MOFs possess a number of advantages over other luminescent materials. Analyte adsorption within MOF pores allows for its pre-concentration, increasing sensor sensitivity. Selectivity in MOFs can be achieved by tuning pore dimensionality and/or by proper functionalization of the linkers. Porosity allows for adsorption of chromophores which can be luminescent themselves or act as antennas, this enlarging the number of mechanisms beyond luminescence. Given my long-lasting interest in the design and synthesis of thiazole-containing linkers and related MOFs/CPs in Firenze (Italy) [13–20], I have personally contributed to this Special Issue with a mini-review article describing the luminescent features of thiazole- and thiadiazole-based MOFs and CPs [21]. Thiazoles are intrinsically fluorescent heterocycles, and their exploitation in the construction of MOFs is still at its infancy. Within this Special Issue, there are two additional contributions on the luminescence topic. Cepeda, Rodríguez-Diéguez et al. (País Vasco/Granada, Spain) published two new Zn^{II} and Cd^{II} CPs based on 1H-indazole-6-carboxylic acid, along with their photoluminescence (PL) properties [22]. Both CPs present emission spectra similar to the free ligand, confirming that the electronic transitions of the polymers are ligand-centered (of π–π^* type). This kind of emission is typical of MOFs and CPs built with an electronically inert metal ion (like Zn^{II} or Cd^{II}, with a d^{10} closed-shell electronic configuration). The Russian team of Artem'ev and colleagues (Novosibirsk, Russian Federation) has synthesized a 1D Ag^I-based CP with 4,6-bis (diphenylphosphino)pyrimidine [23]. The material exhibits pronounced thermochromic luminescence, expressed by reversible changing of the emission chromaticity from yellow (at ambient temperature) to orange (at the liquid nitrogen temperature). The detailed temperature-dependent photophysical study has shown that the ambient temperature photoluminescence may be tentatively ascribed to thermally activated delayed fluorescence (TADF) caused by formation of (M + L′)LCT excited states. This class of materials find promising application in luminescent thermometry.

The water tolerance of MOFs is an essential feature for many practical applications when operating under ordinary humid atmosphere. Resistance toward hydrolysis in such environment is not common, given the intrinsically reactive metal-ligand coordination bond. The most important MOF family that is resistant to hydrolysis is that of the zirconium based "UiOs". Water stability opens greener synthetic routes, and aqueous synthesis of MOFs at room temperature offers many advantages such as reduction in the generation of toxic byproducts and operation costs, as well as increased safety in the material's production. The American team of Islamoglu and Farha (Chicago, US) has reported an aqueous solution-based synthesis of the robust zirconium MOF UiO-66-NO_2 at room temperature [24]. Water vapor sorption isotherms at room temperature indicated high uptake, suggesting the potential of this MOF for adsorption-based cooling applications or water harvesting systems. Water can also be considered a substrate for heterogeneous catalysis. In this regard, the team of Macchioni and Costantino (Perugia, Italy) has functionalized UiO-66 post-synthetically with the organometallic Ir^{III} complex [Ir(HEDTA)Cl]Na, placing it on the defective MOF sites occupied by formate linkers [25]. Anchoring of the complex occurs through an exchange of formate with the free HEDTA carboxylate group. The modified material was tested as a heterogeneous catalyst for the water chemical oxidation reaction (water splitting with concomitant O_2 production) by using cerium ammonium nitrate as sacrificial agent.

In conclusion, I hope that these open-access contributions will serve as guiding lights for future MOFs and CPs development in still undiscovered applications. I thank the authors for their original contributions for the Special Issue, and I thank the reviewers for their tireless enthusiasm, insightful comments and revisions. Last but not least, my warmest acknowledgments go to the *Inorganics* Editorial Staff for its constant dedication, support and patience in collecting so many excellent papers over a time period of two years.

Conflicts of Interest: The author declares no conflict of interest.

References

1. García, H.; Navalón, S. (Eds.) *Metal-Organic Frameworks: Applications in Separations and Catalysis*; Wiley-VCH Verlag GmbH & Co., KGaA.: Weinheim, Germany, 2018.
2. Kaskel, S. (Ed.) *The Chemistry of Metal–Organic Frameworks: Synthesis, Characterization, and Applications*; Wiley-VCH Verlag GmbH & Co. KGaA.: Weinheim, Germany, 2016. [CrossRef]
3. Seyyedi, B. *Metal-Organic Frameworks: A New Class of Crystalline Porous Materials*; Lambert Academic Publishing: Saarbrücken, Germany, 2014.
4. MacGillivray, L.R.; Lukehart, C.M. *Metal-Organic Framework Materials*; John Wiley & Sons: New York, NY, USA, 2014.
5. Farrusseng, D. *Metal-Organic Frameworks: Applications from Catalysis to Gas Storage*; Wiley-VCH Verlag: Weinheim, Germany, 2011.
6. Schroeder, M. *Functional Metal-Organic Frameworks: Gas Storage, Separation and Catalysis*; Springer: Berlin/Heidelberg, Germany, 2010.
7. LEGOÓ. Available online: https://www.lego.com/ (accessed on 29 April 2021).
8. Rancan, M.; Carlotto, A.; Bottaro, G.; Armelao, L. Effect of Coordinating Solvents on the Structure of Cu(II)-4,4′-bipyridine Coordination Polymers. *Inorganics* **2019**, *7*, 103. [CrossRef]
9. Xhaferaj, N.; Tăbăcaru, A.; Moroni, M.; Senchyk, G.A.; Domasevitch, K.V.; Pettinari, C.; Galli, S. New Coordination Polymers of Zinc(II), Copper(II) and Cadmium(II) with 1,3-Bis(1,2,4-triazol-4-yl)adamantane. *Inorganics* **2020**, *8*, 60. [CrossRef]
10. Chen, F.; Drake, H.F.; Feng, L.; Powell, J.A.; Wang, K.-Y.; Yan, T.-H.; Zhou, H.-C. Metal–Organic Frameworks as Versatile Platforms for Organometallic Chemistry. *Inorganics* **2021**, *9*, 27. [CrossRef]
11. Atzori, C.; Ethiraj, J.; Colombo, V.; Bonino, F.; Bordiga, S. Adsorption Properties of $Ce_5(BDC)_{7.5}(DMF)_4$ MOF. *Inorganics* **2020**, *8*, 9. [CrossRef]
12. Taddei, M.; McPherson, M.J.; Gougsa, A.; Lam, J.; Sewell, J.; Andreoli, E. An Optimised Compaction Process for Zr-Fumarate (MOF-801). *Inorganics* **2019**, *7*, 110. [CrossRef]
13. Mercuri, G.; Moroni, M.; Fermi, A.; Bergamini, G.; Galli, S.; Giambastiani, G.; Rossin, A. Zirconium Metal—Organic Frameworks Containing a Biselenophene Linker: Synthesis, Characterization, and Luminescent Properties. *Inorg. Chem.* **2020**, *59*, 15832–15841. [CrossRef]

14. Luconi, L.; Mercuri, G.; Islamoglu, T.; Fermi, A.; Bergamini, G.; Giambastiani, G.; Rossin, A. Benzothiazolium-Functionalized NU-1000: A Versatile Material for Carbon Dioxide Adsorption and Cyanide Luminescence Sensing. *J. Mater. Chem. C* **2020**, *8*, 7492–7500. [CrossRef]
15. Müller, P.; Bucior, B.; Tuci, G.; Luconi, L.; Getzschmann, J.; Kaskel, S.; Snurr, R.Q.; Giambastiani, G.; Rossin, A. Computational Screening, Synthesis and Testing of Metal–Organic Frameworks with a Bithiazole Linker for Carbon Dioxide Capture and its Green Conversion into Cyclic Carbonates. *Mol. Syst. Des. Eng.* **2019**, *4*, 1000–1013. [CrossRef]
16. Staderini, S.; Tuci, G.; Luconi, L.; Müller, P.; Kaskel, S.; Eychmüller, A.; Eichler, F.; Giambastiani, G.; Rossin, A. Zinc Coordination Polymers Containing Isomeric Forms of *p*-(Thiazolyl)benzoic Acid: Blue-Emitting Materials with a Solvatochromic Response to Water. *Eur. J. Inorg. Chem.* **2017**, 4909–4918. [CrossRef]
17. Staderini, S.; Tuci, G.; D'Angelantonio, M.; Manoli, F.; Manet, I.; Giambastiani, G.; Peruzzini, M.; Rossin, A. Zinc Coordination Polymers Containing the *m*-(2-thiazolyl)benzoic Acid Spacer: Synthesis, Characterization and Luminescent Properties in Aqueous Solutions. *ChemistrySelect* **2016**, *6*, 1123–1131. [CrossRef]
18. Tuci, G.; Giambastiani, G.; Kwon, S.; Stair, P.C.; Snurr, R.Q.; Rossin, A. Chiral Co(II) Metal−Organic Framework in the Heterogeneous Catalytic Oxidation of Alkenes under Aerobic and Anaerobic Conditions. *ACS Catal.* **2014**, *4*, 1032–1039. [CrossRef]
19. Rossin, A.; Tuci, G.; Giambastiani, G.; Peruzzini, M. 1D and 2D Thiazole-Based Copper(II) Coordination Polymers: Synthesis and Applications in Carbon Dioxide Capture. *ChemPlusChem* **2014**, *79*, 406–412. [CrossRef] [PubMed]
20. Rossin, A.; Di Credico, B.; Giambastiani, G.; Peruzzini, A.; Pescitelli, G.; Reginato, G.; Borfecchia, E.; Gianolio, D.; Lamberti, C.; Bordiga, S. Synthesis, Characterization and CO_2 Uptake of a Chiral Co(II) Metal–Organic Framework Containing a Thiazolidine-Based Spacer. *J. Mater. Chem.* **2012**, *22*, 10335–10344. [CrossRef]
21. Mercuri, G.; Giambastiani, G.; Rossin, A. Thiazole- and Thiadiazole-Based Metal–Organic Frameworks and Coordination Polymers for Luminescent Applications. *Inorganics* **2019**, *7*, 144. [CrossRef]
22. García-Valdivia, A.A.; Echenique-Errandonea, E.; Ramírez-Rodríguez, G.B.; Delgado-López, J.M.; Fernández, B.; Rojas, S.; Cepeda, J.; Rodríguez-Diéguez, A. Photoluminescent Coordination Polymers Based on Group 12 Metals and 1H-Indazole-6-Carboxylic Acid. *Inorganics* **2021**, *9*, 20. [CrossRef]
23. Artem'ev, A.V.; Davydova, M.P.; Berezin, A.S.; Samsonenko, D.G. Synthesis and Thermochromic Luminescence of Ag(I) Complexes Based on 4,6-Bis(diphenylphosphino)-Pyrimidine. *Inorganics* **2020**, *8*, 46. [CrossRef]
24. Chen, Z.; Wang, X.; Islamoglu, T.; Farha, O.K. Green Synthesis of a Functionalized Zirconium-Based Metal–Organic Framework for Water and Ethanol Adsorption. *Inorganics* **2019**, *7*, 56. [CrossRef]
25. Gatto, G.; Macchioni, A.; Bondi, R.; Marmottini, F.; Costantino, F. Post Synthetic Defect Engineering of UiO-66 Metal–Organic Framework with An Iridium(III)-HEDTA Complex and Application in Water Oxidation Catalysis. *Inorganics* **2019**, *7*, 123. [CrossRef]

inorganics

MDPI

Review

Thiazole- and Thiadiazole-Based Metal–Organic Frameworks and Coordination Polymers for Luminescent Applications

Giorgio Mercuri [1], Giuliano Giambastiani [1,2] and Andrea Rossin [1,*]

[1] Istituto di Chimica dei Composti Organometallici (ICCOM-CNR), Via Madonna del Piano 10, 50019 Sesto Fiorentino, Italy; giorgio.mercuri@iccom.cnr.it (G.M.); giuliano.giambastiani@iccom.cnr.it (G.G.)
[2] Institute of Chemistry and Processes for Energy, Environment and Health (ICPEES), UMR 7515 CNRS University of Strasbourg (UdS), 25, rue Becquerel, 67087 Strasbourg CEDEX 02, France
* Correspondence: a.rossin@iccom.cnr.it; Tel.: +39-055-522-5248

Received: 15 November 2019; Accepted: 12 December 2019; Published: 14 December 2019

Abstract: This mini-review focuses on the 2015–2019 literature survey of thiazole- and thiadiazole-containing Metal–Organic Frameworks (MOFs) and Coordination Polymers (CPs) exploited in the applicative field of luminescent sensing.

Keywords: thiazole; thiadiazole; metal–organic frameworks; coordination polymers; luminescence

1. Introduction

Chemical sensors, i.e., analytical tools that provide information about the chemical composition of the environment in which they are introduced, are based on a variety of transduction mechanisms (e.g., optic, electronic, optoelectronic). Luminescent materials release energy in the form of electromagnetic radiation in the visible region in response to external stimuli [1]. Recently, luminescent materials have gained the stage as chemical sensors [2], as luminescence is among the most desirable transduction mechanisms for its relative easiness of use, technical simplicity and broad adaptability [3].

Metal–organic frameworks (MOFs) are hybrid organic/inorganic 3D coordination polymers with open structures deriving from the self-assembly of poly(topic) ligands and metal ions or metal-based clusters [4]. In the past fifteen years, MOFs have emerged as a promising alternative to all inorganic materials (e.g., activated carbons and zeolites) in industrially and technologically relevant applications requiring porous compounds. As chemical sensors, luminescent MOFs possess a number of advantages over other luminescent materials. Analyte adsorption within MOF pores allows for its pre-concentration, increasing sensor sensitivity. Selectivity in MOFs can be achieved by tuning pore dimension and/or by proper functionalization of the linkers. Porosity allows for adsorption of chromophores which can be luminescent themselves or act as antennas, this enlarging the number of mechanisms beyond luminescence [5]. MOFs structure flexibility may lead to changes in the local coordination environment, hence in the emissive properties, upon guest adsorption [6]. Compared to amorphous luminescent compounds, MOFs possess well-defined crystal structures, which allows theoretical and experimental study of the sensing mechanisms at the molecular level. Finally, they are less expensive than other detection methods.

MOFs luminescence may have different origins [7–10]: ligand-based, metal-based, charge-transfer, or guest-induced (Figure 1). Linker-based luminescence is commonly found in MOFs containing conjugated ligands with extended π-delocalized carbo- or heterocycles that absorb in the visible or UV region, combined with metallic nodes made of metal ions in a d^{10} or d^{0} electronic configuration, such as Zn(II), Cd(II) or Zr(IV) (i.e., largely electronically inert ions) [10–12]. Metal-centered luminescence (phosphorescence) requires the presence of open-shell configuration metal ions. Therefore, it is mostly seen in MOFs containing f-elements as nodes [13], although some examples with d^{10} metal ions are also present in the literature [14]. Fluorescent ligands typically bear π-conjugated electrons which can give emissions upon irradiation. Signal transduction in luminescent MOFs normally takes place through emission quenching (more occasionally enhancement) after a guest adsorption. The strength of these effects depend on the nature of the host–guest interactions. In the most interesting cases, adsorbates induce a shift in the emission wavelength of the MOF emission. This type of signal transduction is inherently more attractive than simple quenching. If the emission band falls in the visible region, the marked evidence of the occurred guest capture is immediately observable as a color change at the naked eye.

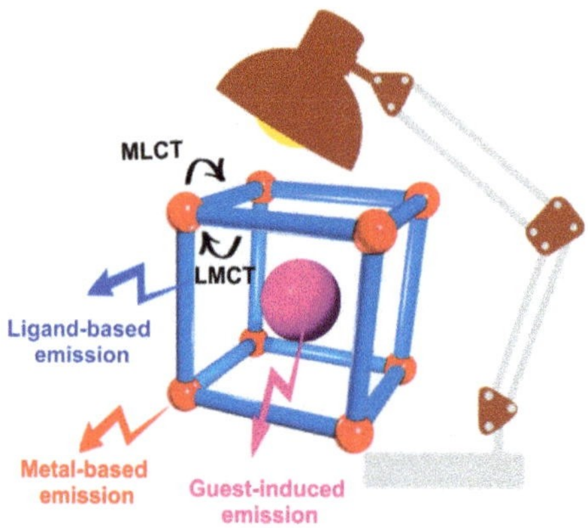

Figure 1. Pictorial representation of the mechanisms at the origin of MOFs luminescence.

More recently, the discovery and development of MIXMOFs (also known as Multivariate MOFs) [15–17] has opened new horizons in chemical sensing. The simultaneous presence of different linkers within a unique solid phase is useful for multiple sensing capability (multiplexing), i.e., the detection of different analytes with the same material combined with an extremely high sensitivity (low detection limits, Figure 2).

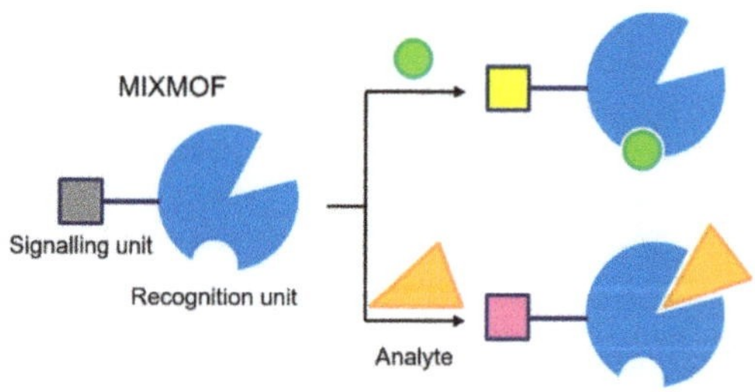

Figure 2. Pictorial representation of multiple sensing by MIXMOFs.

As for the target analytes of interest, MOFs have been proven to be efficient chemical sensors for gases [18–23], small molecules like solvents [24–27], amines [28,29], volatile organic compounds (VOCs) [30–32], explosives (nitroarenes) [33–35], bioactive compounds [36,37] and dyes [6]. Finally, detection of polluting metal ions (both cations and anions) in aqueous solutions is also a widely exploited application field [38–46].

The typical MOFs ligands belong to the classes of (aliphatic or aromatic) polycarboxylic acids and polydentate N-heterocyclic bases (such as pyridines, imidazoles, pyrazoles and tetrazoles) [47,48].

The presence in the heterocycle of atoms different from nitrogen is relatively rare. In fact, there are much fewer examples of MOFs or CPs built with heterocycles containing oxygen, sulfur or selenium in their skeleton. Heteroatoms with pronounced differences in their relative Lewis basicity modify the electron density distribution within the linker and may show different coordination ability for metal ions used in the MOF synthesis when seen as alternative coordination sites. In addition, the abundance of electron-rich atoms in the linkers increases the number of weak non-covalent interactions (like hydrogen bonds or π–π stacking) in the solid state, providing more thermally/chemically stable materials and generating fascinating crystalline scaffolds. Within the applicative context of luminescent sensing, organic light emitting diodes (OLEDs) and organic optoelectronics, π-conjugated heterocyclic compounds containing a sulfur atom like thiophene exhibit an extensive variety of optical, electrical and photoelectric applications. The introduction of an additional nitrogen atom in the linker should improve these features considerably. Consequently, in recent years there has been a growing interest in (N,S) containing heterocycles. Among the simplest and most naturally occurring (N,S) heterocycles there are thiazoles and thiadiazoles (Scheme 1). Thiazoles are intrinsically fluorescent [49,50] and they can be found in several naturally occurring biomolecules like vitamin B1 (thiamine) or luciferin (the active component generating luminescence in fireflies). They have been exploited for the preparation of synthetic fungicides like thiabendazole, anti-inflammatory drugs like Meloxicam or in the fabrication of OLEDs and semiconductors [51,52]. More recently, their application as fully organic sensitizers in solar cells (DSSC) for energy applications has also started to appear in the literature [53,54].

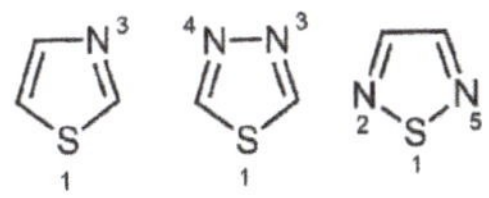

Scheme 1. 1,3-thiazole, 1,3,4-thiadiazole and 1,2,5-thiadiazole.

The preparation of thiazole-based MOFs/CPs (from ligand design to the polymer construction) is one of our main research activities [55–59]. In the last five years, there has been a significant boost in the interest on this specific topic. Following this interest and as an extension of a Highlight article published by our group in 2015 [60], in here we would like to present a 2015–2019 survey of thiazole- and thiadiazole-containing MOFs and CPs for applications in luminescence. The mini-review will be divided into two chapters: thiazole-based MOFs/CPs (Section 2) and thiadiazole-based MOFs/CPs (Section 3). We believe that this collection may be useful for other research groups who are actively working on the same research field worldwide as a reference work to find new inspiration and ideas for a better development of this intriguing and promising applicative field.

2. Thiazole-Based MOFs and CPs as Luminescent Materials

In the following Sections the literature survey is presented and classified according to both ligand type and publication year, moving from the oldest (2015) to the most recent (2019) examples within each category. A collective thiazole-based ligands scheme (Scheme 2) is reported here for the sake of clarity.

Thiazole, as such, has been exploited as a solvent in the synthesis of a series of lanthanide CPs from anhydrous $LnCl_3$ and 1,2-bis(4-pyridyl)ethylene (dpe) under solvothermal conditions by the group of Müller-Buschbaum (Wurzburg, DEU) [61]. The obtained polymers of general formula $[LnCl_3(dpe)(thiazole)_2]_\infty$ are 1D or 2D, and the connections between adjacent metal ions in the crystal structures are guaranteed by the dpe bridging linker. Thiazole always acts as an end-on (terminal) N-bound ligand. Dpe was proved to be a suitable sensitizer for the photoluminescence of lanthanides in the near-infrared region (NIR) only: ligand-sensitized 4*f*–4*f* NIR emission is dominating for Nd(III), Er(III) and Yb(III). The same group has subsequently prepared the analogues containing the "saturated counterpart" of dpe, namely 1,2-bis(4-pyridyl)ethane (bpe), using the same synthetic approach [62]. Depending on the ligand content, assorted structures from 3D frameworks $\{[LnCl_3(bpe)_2]\cdot thiazole\}_\infty$

(Ln = Ce–Lu), to the 1D-strand $[La_2Cl_6(bpe)_2(thiazole)_6]_\infty$ were obtained. The compounds exhibit a variety of luminescence properties and different phenomena. These included ligand-centered fluorescence, metal-centered *5d–4f*/*4f–4f* emission in the visible and the NIR range, and antenna effects via Dexter and Förster (FRET) energy transfer mechanisms (non-radiative processes with electron exchange). The obtained results proved that the linker bpe is also an excellent sensitizer for lanthanide emission, as observed for dpe. Finally, in 2017 the same research team replaced the pyridyl-based linkers with other N-containing heterocycles: pyrazine (pyz) and pyrimidine (pym) [63], forming the compounds $[Ln_2Cl_6(pyz)(thiazole)_6]_\infty$ and $[Ln_2Cl_6(pym)_2(thiazole)_4]_\infty$ (Ln = Tb, Er). The polymers with Tb and Er show intense luminescence with a strong sensitizer effect of the ligands that can be observed by dominant ligand excitation and exclusive metal emission for VIS and NIR.

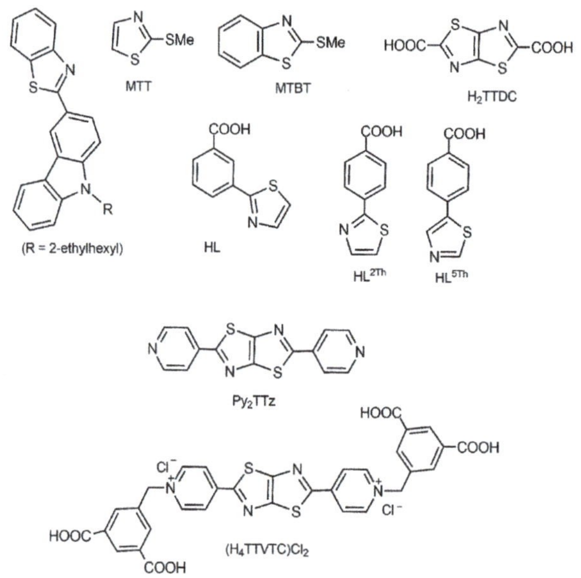

Scheme 2. Overview of the structures of the thiazole-containing ligands discussed in Section 2.

The group of Tang, Wang and co-workers (Kumming, CHN) prepared a novel red-emitting cationic iridium(III) coordination polymer using the thiazole-based ligand 2-(9-(2-ethylhexyl)-9H-carbazol-3-yl)benzo [d]thiazole (Scheme 2) and 4,4′-bipyridine (bipy). It can be efficiently excited by blue light of GaN chips when blended in YAG:Ce-GaN-based cold white LEDs at 0.1 wt % and 0.2 wt %. The emission became neutral white light, and the collected results suggested that the coordination polymer is a promising red-emitting phosphor candidate for warm white LEDs [64].

Artem'ev and collaborators (Novosibirsk, RUS) exploited 2-(methylthio)thiazole (MTT) and 2-(methylthio)benzothiazole (MTBT) (Scheme 2) to prepare the corresponding copper(I) 1D CPs $[CuI(MTT)]_\infty$ and $[Cu_2I_2(MTBT)_2]_\infty$, respectively [65]. The former consists of $(–Cu–I–Cu–I–)_\infty$ zigzag chains wherein the adjacent Cu atoms are bridged by the MTT ligand through the thiazole N and sulfide S atoms. In the latter, a planar $[Cu(\mu\text{-}I)]_2$ dimeric core is supported by the two ligands in a N-monodentate manner. At ambient temperature, the former exhibits a weak yellow luminescence (λmax = 570 nm), while the latter emits in the deep-red region (λmax = 705 nm), thus representing a rare example of a Cu(I) coordination compound showing an emission in this color range. One year later, the same group started from the same ligands to prepare the Ag(I) analogues $[Ag(MTT)(NO_3)]_\infty$ and $[Ag_2(MTBT)_2(NO_3)_2(MeCN)_2]_\infty$ as layered 2D CPs exhibiting Ag···Ag interactions in the solid state [66]. The latter demonstrates dual emission behavior associated with simultaneous presence of high-energy and low-energy emission bands originated by Ag(I)-perturbed singlet and triplet π–π* and n–π* intraligand transitions, respectively.

In recent years, our group has reported the preparation of Zn(II) CPs with different isomeric forms of (thiazolyl)benzoic acid. Four 1D coordination polymers have been prepared, starting from the organic linker *m*-(2-thiazolyl)benzoic acid (HL, Scheme 2), also combined with auxiliary ligands: $[Zn_2(L)_4(H_2O)\cdot 2(MeCN)]\infty$, $[Zn_2(L)_3(NO_3)(bipy)]\infty$, $[Zn_2(L)_4(bipy)]\infty$, and $[Zn_2(L)_4(bpe)]\infty$. In all species, the Zn_2(carboxylate)$_4$ "paddle-wheel" dimer is the constituting inorganic node, where the carboxylate groups from L^- are bridging two adjacent metal centers [56]. The polymeric nature stems from the auxiliary N-donors that are shared by adjacent Zn_2 dimers. The luminescent properties in aqueous solutions of $[Zn_2(L)_4(bipy)]\infty$ and $[Zn_2(L)_4(bpe)]\infty$ have been examined. The N and S donors dangling from the thiazole rings in these polymers can engage into further supramolecular interactions with (acidic) metal cations, inducing a luminescence quenching after complexation. Their photophysical properties have been studied, in the presence and in the absence of Hg(II) ions. As expected, the electronic transitions observed are mainly ligand-centered, given the "closed shell" electronic configuration of the Zn(II) ion (d^{10}) that bans any metal-to-ligand or ligand-to-metal charge transfer transitions. The complexation of an acidic cation like Hg(II) to the basic sites (N) dangling from the ligand side-arms induces a luminescence quenching in both absorption and emission (Figure 3), along with a partial polymer aggregation in solution (as judged from the Dynamic Light Scattering results).

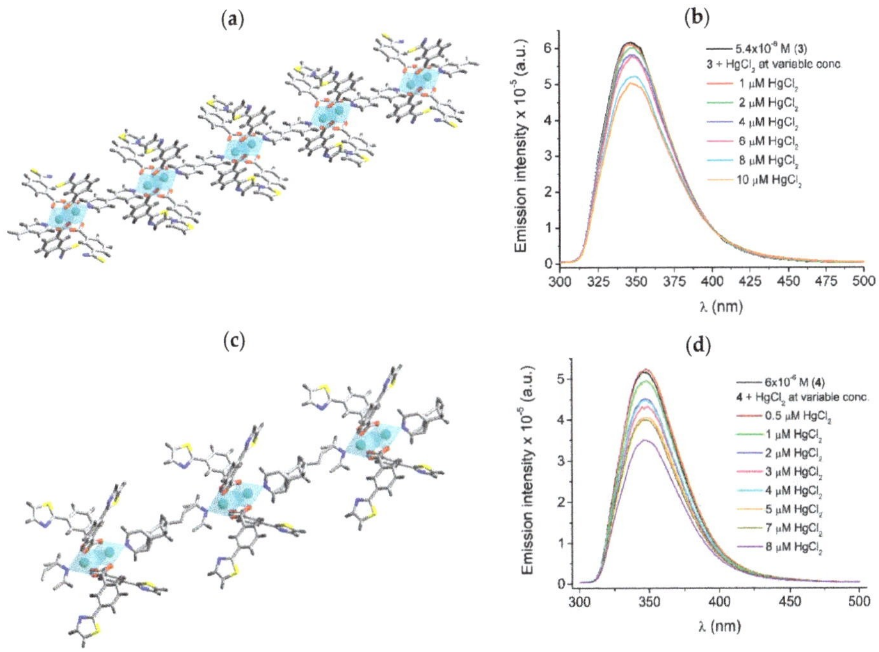

Figure 3. The 1D chain in the lattice of $[Zn_2(L)_4(bipy)]\infty$ (**a**) and $[Zn_2(L)_4(bpe)]\infty$ (**c**), along with the related emission spectra at variable Hg(II) concentrations in aqueous solutions (**b**,**d**). Taken from reference [56]. Reproduced with permission from Wiley VCH.

One year later, we moved to the home-made isomers *p*-(2-thiazolyl)benzoic acid (HL^{2Th}) and *p*-(5-thiazolyl)benzoic acid (HL^{5Th}, Scheme 2). Two coordination polymers of assorted dimensionality (1D, 2D) have been prepared, namely $[Zn_3(L^{2Th})_4(OH)_2\cdot 2(HL^{2Th})]\infty$ and $[Zn(L^{5Th})(OAc)]\infty$, respectively [55]. The luminescent properties of both polymers have been analyzed in the solid state; they feature ligand-centered emissions at λ = 434 nm and λ = 427 nm. These electronic transitions fall in the visible region, giving the samples a characteristic blue color under an ordinary UV lamp. The theoretical analysis of the electronic features of the ligands and related molecular orbitals reveals that the observed transitions are mainly of $\pi \rightarrow \pi^*$ nature, involving π orbitals delocalized on both aromatic cycles. A significant (reversible) blue-shift of the emission maximum of ca. 60 nm from the visible to the UV region was observed for the 1D polymer (where the N-thiazole donors are dangling from the polymeric chain) when suspended in water (Figure 4).

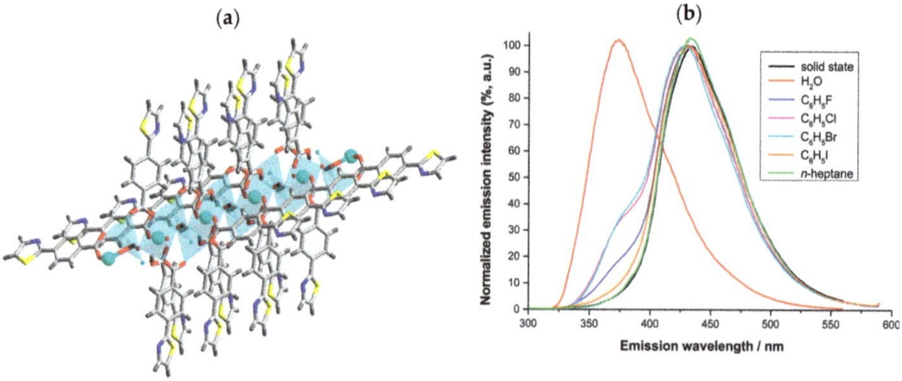

Figure 4. The 1D chain in the lattice of $[Zn_3(L^{2Th})_4(OH)_2]\infty$ (**a**) and its emission spectra when suspended in solvents with different polarity (**b**). Taken from reference [55]. Reproduced with permission from Wiley VCH.

Starting from 2018, the investigation on this field focused on a highly luminescent structural core: thiazolo [5,4-*d*]thiazole (H_2TTDC), made of two fused 1,3-thiazole rings along the C(4)–C(5) bond (Scheme 2). This chemical motif is planar, rigid and conjugated, thus representing an optimal fluorescent molecule. Indeed, it has been widely exploited in applications dealing with organic electronics and solar cells because of its appealing features [67–69]. Thiazolo [5,4-*d*]thiazole can be easily functionalized on C(2) and C(7) with coordinating groups suitable for the construction of MOFs and CPs. The group of Falcao et al. (Pernambuco, BRA) exploited thiazolo [5,4-*d*]thiazole-2,7-dicarboxylate ($TTDC^{2-}$, Scheme 2) for the construction of lanthanide two-dimensional CPs of general formula $[Ln_2(TTDC)_3 8H_2O]\infty$ (Ln = La, Ce, Nd, Sm, Eu, Gd) [70]. In these polymers, the rare Earths possess coordination number 9, and the $TTDC^{2-}$ ligand adopts different coordination modes, either terminal monodentate or (*N*,*O*)-chelating bidentate through one N atom from the thiazole ring. This kind of behavior is frequently found in derivatives of thiazole-2- and thiazole-4-carboxylic acids [60]. Luminescence spectra for the Nd(III), Sm(III) and Eu(III) structures show the typical transitions of these ions, through the antenna effect promoted by the ligand. Since the simple thiazolo [5,4-*d*]thiazole-2,7-dicarboxylate is thermally labile, losing the carboxylic groups as CO_2 at temperatures as low as 70 °C, $TTDC^{2-}$ is not an ideal linker for solvothermal MOFs/CPs syntheses. Thus, alternative coordinating groups have been later attached to the end-carbon atoms of thiazolo [5,4-*d*]thiazole. The most widely exploited molecule so far is 2,7-bis(4-pyridyl)thiazolo [5,4-d]thiazole (Py2TTz, Scheme 2). Du, Zang and co-workers (Zhengzhou, CHN) prepared mixed-ligand MOFs (MIXMOFs) combining Zn(II) or Cd(II) with Py_2TTz and terephtalic acid, obtaining pillared cubic structures of pcu topology: $[M(Py_2TTz)(BDC)\cdot 2(DMF)]\infty$, where M = Zn, Cd; DMF = *N*,*N*-dimethylformamide and BDC^{2-} = terephtalate[71]. The pillar-layered frameworks consist of 2D sheets [constructed by ligand BDC^{2-} and "paddlewheel" secondary building units $M_2(COO)_4$] and Py_2TTz that pillars neighboring sheets. The frameworks are twofold-interpenetrated because of the long linker. Their fluorescence properties were investigated systematically. The results show that both MOFs display good fluorescent properties, which can be efficiently quenched by a trace amount of nitroaromatics 2,4,6-trinitrophenol and antibiotic nitrofurazone in water. The small limit of detection demonstrates that they can serve as good fluorescent sensors for trinitrophenol and nitrofurazone in aqueous environment. One year later, the same group has reported on the preparation of structural analogues containing variously substituted terephtalates and also with naphthalene-2,6-dicarboxylic acid (H_2NDC) [72]. The presence of a linear dicarboxylate linker like BDC^{2-} (with the coordinating groups at 180° with respect to each other) favors the formation of a multi-fold interpenetrated crystal lattice, while a V-shaped ligand like NDC^{2-} affords a 2D layered structure. All the synthesized compounds demonstrate structure-dependent luminescence in the solid state at room temperature, which are different from those of the pristine ligands. Compared to the emission peaks of the free ligands, the emission spectra of the MOF samples demonstrate different degrees of red shift. The nine MOFs prepared display structure-dependent emissions ranging from blue to blue-white and blue-green (Figure 5).

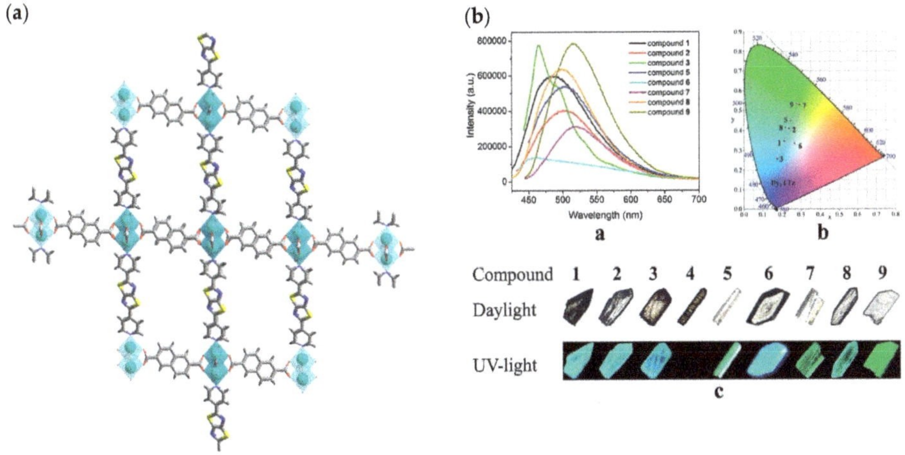

Figure 5. (**a**) Crystal structure of $[Zn_2(Py_2TTz)(NDC)_2]\infty$; (**b**) Solid-state emission spectra, CIE chromaticity coordinates and crystal images under daylight and ultraviolet light of the family of compounds reported in Reference [72]. Reproduced with permission from Wiley VCH.

The group of Saha and co-workers (Clemson, USA) has studied the luminescent behavior of $[Zn_2(Py_2TTz)(NDC)_2]\infty$ as such and in the presence of assorted transition metal cations [73]. NDC^{2-} struts serve as antenna chromophores and energy donors, while Py_2TTz pillars act as complementary energy acceptors and light emitters. The highly ordered crystalline arrangement and the good overlap between their emission and absorption spectra enabled ligand-to-ligand Förster resonance energy transfer, allowing the MOF to display exclusively Py_2TTz-centric blue emission at $\lambda = 410$ nm. In the presence of Hg(II), the photoluminescence of the MOF underwent a significant red shift to 450 nm followed by quenching, whereas other transition metal ions [Mn(II), Fe(II), Co(II), Ni(II), Cu(II) and Cd(II)] caused only fluorescence quenching but no shift. Consequently, $[Zn_2(Py_2TTz)(NDC)_2]\infty$ can be considered an efficient Hg(II) luminescent sensor.

The group of Arici et al. (Eskiseir, TUR) prepared three novel Co(II) coordination polymers, formulated as $\{[Co(\mu_3\text{-ipa})(Py_2TTz)]\cdot DMF\}\infty$, $\{[Co(\mu_3\text{-fdc})(Py_2TTz)(H_2O)]\cdot DMF\}\infty$ and $\{[Co(\mu_3\text{-tdc})(Py_2TTz)]\cdot G\}\infty$ (ipa^{2-} = isophtalate, fdc^{2-} = furan-2,5-dicarboxylate, tdc^{2-} = thiophene-2,5-dicarboxylate, G = guest solvent molecule) [74]. When the angles between the two carboxylate groups increase, the dimensionality of the compounds increases. Indeed, compound $\{[Co(\mu_3\text{-ipa})(Py_2TTz)]\cdot DMF\}\infty$ (where the $-COO^-$ groups lie at 120° with respect to each other) has a 2D layered structure, while the others (where the angle between the two carboxylates is around 145°)

are 3D frameworks with pcu topology. The absorption bands of the polymers are similar to those of the corresponding organic constituents, mainly of π–π^* nature.

In an intriguing variation on the theme, Gao et al. (Shanghai, CHN) designed an "extended viologen" organic linker through quaternization of the pyridine N atoms in Py_2TTz, forming the new tetracarboxylic acid 2,7-bis(1-(3,5-dicarboxylatobenzyl)pyridin-1-ium-4-yl)-thiazolo [5,4-d]thiazole ($TTVTC^{2-}$, Scheme 2) [75]. Its incorporation into the Zn(II) MOF $\{[Zn(TTVTC)]\cdot 4H_2O\}\infty$ produced a highly water-stable material that acts as an efficient and recyclable sensor for the detection of Cr(VI) in aqueous solutions. Extended viologen compounds have been shown to be strongly fluorescent; indeed, $[H_4TTVTC]Cl_2$ shows strong cyan emission under UV light (365 nm), and the solid-state fluorescence spectrum shows an emission band centered at 485 nm. The related Zn-MOF also shows strong fluorescence, the color becoming yellow-green and the emission maximum shifting to 524 nm. The fluorescence of the aqueous dispersions was almost completely quenched by K_2CrO_4 and $K_2Cr_2O_7$. By contrast, many other anions at the same concentration did not have significant effects (Figure 6).

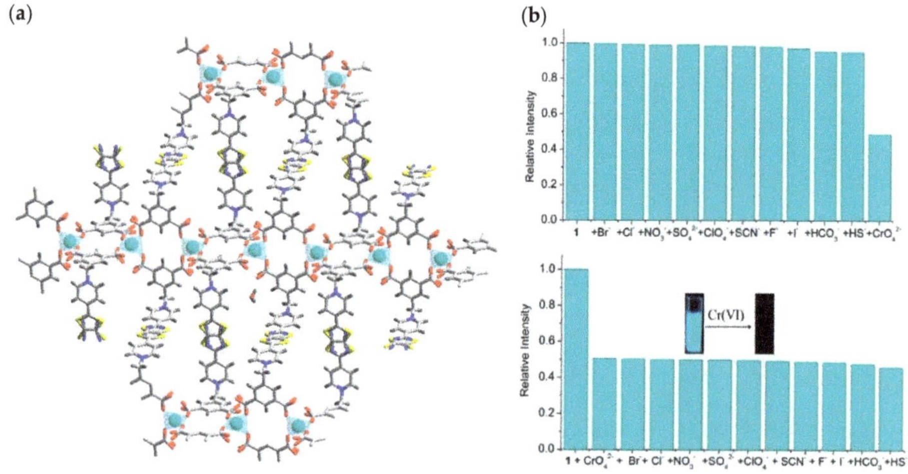

Figure 6. (**a**) Crystal structure of $\{[Zn(TTVTC)]\cdot 4H_2O\}\infty$; (**b**) Its relative luminescent intensity by successively adding anions. The inset shows the visual fluorescence quenching by CrO_4^{2-} or $Cr_2O_7^{2-}$. Adapted with permission from Reference [75]. Copyright 2019. American Chemical Society.

3. Thiadiazole-Based MOFs and CPs as Luminescent Sensors

For thiadiazole, the two different isomers 1,3,4-thiadiazole and 1,2,5-thiadiazole will be discussed together within this Section. As for thiazole, Scheme 3 collects the chemical structures of the thiadiazole-based organic ligands of this paragraph.

In 2015, Wang and collaborators (Urumqi, CHN) reported on the synthesis of Cu(II) and Cd(II) CPs with different dimensionality (either 1D chains or 2D sheets) built with the 1,3,4-thiadiazole-containing isomeric linkers L^1, L^2 and L^3 (Scheme 3) [76]. The emission spectra of L^3 and its Cd(II) 2D polymer $[Cd_2(L_3)_3(NO_3)_4(H_2O)_2]\infty$ in the solid state at room temperature were investigated. L^3 exhibits a weak blue fluorescent emission band around 410 nm upon excitation at 350 nm, while the Cd(II) polymer exhibits an intense blue fluorescent emission band around 465 nm.

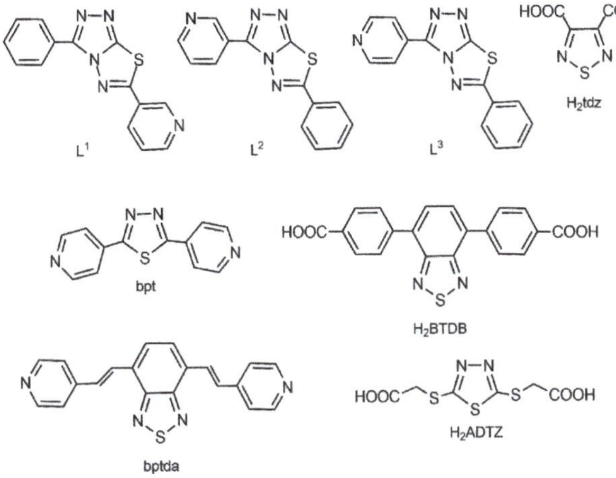

Scheme 3. Overview of the structures of the thiadiazole-containing ligands discussed in Section 3.

The insertion of two 4-pyridyl units on C(2) and C(5) of 1,3,4-thiadiazole gives the neutral bridging linker bpt = 2,5-bis(4-pyridyl)-1,3,4-thiadiazole (Scheme 3). This spacer has been exploited in 2016 by Ma and co-workers (Luoyang, CHN) for the construction of a Cd(II) MOF with general formula $[Cd_2(pam)_2(bpt)_2(H_2O)]_\infty$·($pam^{2-}$ = pamoic acid, a dimer of naphthalenecarboxylic acid) [77]. Its fluorescent properties in the solid state indicate that this MOF may be a potential green-light-emitting material. In fact, the MOF gives a sharp fluorescent emission peak in the green region with a maximum at λ = 517 nm, displaying a large red shift relative to those of free ligands (472 nm and 385 nm for pamoic acid and bpt, respectively). More recently, the same bpt spacer combined with 5-aminoisophtalic acid (H_2aiph) has been used by the team of Bi, Fan and co-workers (Qingdao, CHN) to build a Ni(II) MOF with general formula $[Ni(aiph)(bpt)(H_2O)]_\infty$ [78]. Rhodamine B, methylene violet and methylene blue are the most common organic dyes in waste water. In this work, the nickel MOF was used to degrade aqueous solution containing these dyes under UV irradiation. After 90 min of irradiation, 89.5% of methylene blue, 84.7% of methylene violet and 76.4% of rhodamine B were decomposed in the presence of the MOF, showing an excellent photocatalytic performance in pollutants degradation, even if it has been found that the initial concentration and the catalysts loading both significantly influenced the degradation efficiency.

A new, strongly luminescent Zr(IV)-based metal–organic framework (MOF) material having a UiO-68 (UiO = University of Oslo) framework topology and incorporating the π-conjugated, thiadiazole-functionalized H_2BTDB ligand (H_2BTDB = 4,4′-(benzo[c][1,2,5]thiadiazole-4,7-diyl)dibenzoic acid, Scheme 3) was synthesized under solvothermal conditions in DMF by the group of Biswas (Assam, IND) [79]. The MOF has minimal formula $[Zr_6O_4(OH)_4(BTDB)_6]\cdot 8H_2O\cdot 6DMF$. It showed an emission band centred at 510 nm upon excitation at 370 nm, assigned to a π–π* intraligand transition. Its emission band is slightly blue-shifted and the fluorescence intensity enhanced by 69%, as compared to the pristine organic ligand. The thermally activated form of the compound (after crystallization solvent removal) showed a selective sensing behavior towards 2,4,6-trinitrophenol (commonly known as picric acid), even in the presence of other potentially competing nitroaromatic explosive compounds. The highest fluorescence quenching ability of trinitrophenol can be attributed to both energy and electron transfer processes as well as electrostatic interactions between the hydroxyl group of trinitrophenol and the Lewis basic N-donor sites of the thidiazole-based ligand. In 2019, the same MOF has also been exploited by Eddaoudi and co-workers (KAUST, SAU) for the detection of amines in ultra-low traces (via fluorescence turn-on) in aqueous solutions [80]. This extremely powerful action is driven by hydrogen bonding interactions between the linker and the hosted amines. This observation is supported by density functional theory (DFT) calculations, which clearly corroborate the suppression of the twisting motion of the benzo[c][1,2,5]thiadiazole core in the presence of the amine guest, reducing significantly the non-radiative recombination pathways and subsequently, enhancing the emission intensity (Figure 7).

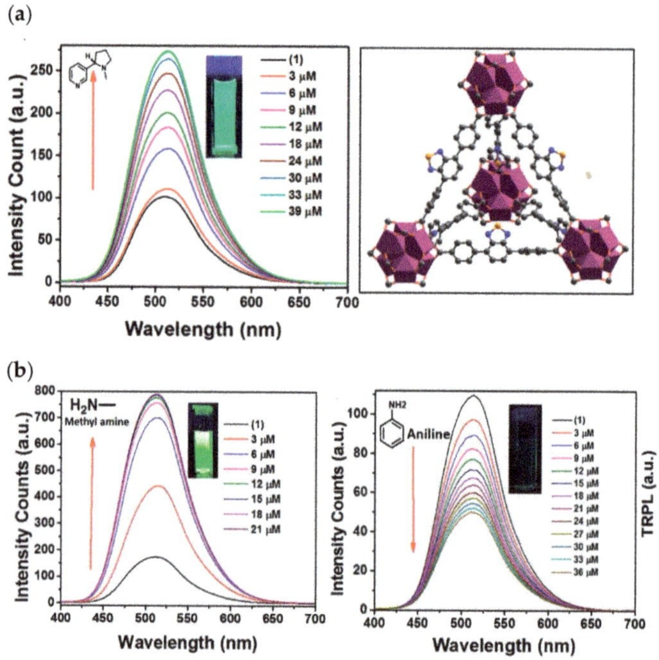

Figure 7. (**a**) The increase in fluorescence of $[Zr_6O_4(OH)_4(BTDB)_6]\infty$ upon addition of 3 μM nicotine, and its chemical structure. (**b**) Fluorescence intensity of an aqueous suspension of the MOF upon addition of 3 μM of methylamine or aniline (λ_{max} = 515 nm). Reproduced with permission from Reference [80]. Copyright 2019. American Chemical Society.

The same benzo[c][1,2,5]thiadiazole central motif decorated by different coordinating side-arms has been prepared by Yang and Zhao (Tianjin, CHN). The coordinating ends are now 4-pyridyl groups, in the tailor-made linker 4,7-bis((*E*)-2-(pyridin-4-yl)vinyl)benzo [2,1,3]thiadiazole (bptda, Scheme 3) [81]. This ligand, in diverse stereochemical conformations around the C=C double bonds present on its side-arms, has been combined with assorted dicarboxylic acids of variable bend angles to create four different Zn(II) MOFs. The main emission peak of the ligand, and also of the related MOFs, fall around λ = 560 nm. The functionalization effect of thiadiazole groups on the optical absorption and visible-light-driven photocatalytic degradation of rhodamine B, methylene blue and crystal violet has been investigated, showing promising applications as dye degradation materials.

A 2D Cd(II) coordination polymer containing the 2,5-(s-acetic acid)dimercapto-1,3,4-thiadiazole (H_2ADTZ, Scheme 3) ligand has been published by the group of Wang (Jinzhou, CHN): $[Cd(bmbpd)_{0.5}(ADTZ)(H_2O)]\infty$, where bmbpd = *N*,*N*′-bis(4-methylenepyridin-4-yl)-1,4-benzenedicarboxamide [82]. Its photoluminescent behavior in the presence of assorted organic solvents showed that it can be a good luminescent sensor (through emission quenching) for ethanol; furthermore, among the scrutinized samples it showed the best catalytic activity in methylene blue photodegradation (60%).

The heterocyclic fluorogenic linker 1,2,5-thiadiazole-3,4-dicarboxylate (H_2tdz, Scheme 3) offering diverse coordination modes has been utilized to synthesize two 3D metal organic frameworks, [Cd(tdz)(bipy)]∞ and [Cu(tdz)(bipy)]∞, by the research team of Mandal (Mohali, IND) [83]. Owing to its excellent stability in water, the photoluminescence spectrum of the Cd(II) polymer was studied in water. It exhibits a very strong emission intensity and is therefore considered a suitable material for sensing studies. The exposed nitrogen atoms in the tdz^{2-} linker are the preferential interaction sites with the hydroxy-containing nitro-aromatic explosives 2,4,6-trinitrophenol, 2,4-dinitrophenol and 4-nitrophenol. Indeed, the trend of increasing quenching efficiency (96% for the trinitrophenol and ~90% for the others) is observed in accordance with the increase in the acidity of the hydroxyl proton of the nitro-explosives (in turn related to the increase in the number of electron-withdrawing nitro groups).

4. Conclusions and Perspectives

It is well known that luminescent MOFs show great promise as various types of sensors owing to their remarkable structural diversity and tunable luminescent properties. This field of investigation is still open to new contributions in terms of ligand design and MOF construction. Generally speaking, linkers containing multiple exposed basic sites (N atoms of heterocycles in particular) are particularly useful to prepare porous luminescent MOFs with high sensing capability towards guests with protic groups, through the switch of N···H hydrogen bonding interactions. In the specific case of thiazoles and thiadiazoles, the presence of the sulfur atom is also useful to promote soft acid-soft base interactions with heavy metal polluting cations like Hg(II), Cd(II), Pb(II), thus creating optimal luminescent sensors for these pollutants in aqueous solutions. A future research direction in this area may be represented by the tailored synthesis of polycyclic rigid aromatic linkers containing more than one thiazole and/or thiadiazole rings. Subsequently, the preparation of highly porous and water-stable MOFs with these linkers is mandatory, representing the second step towards the obtainment of an optimal solid-state sensor. The main hurdle in this synthetic step is represented by the very low linker solubility in the solvent used for the MOF synthesis (normally polar and high-boiling solvents like DMF or *N,N*-diethylformamde (DEF)). The higher the linker conjugation degree and molecular weight, the lower the solubility. A promising approach to solve this problem is represented by mechanochemical (solvent-free) synthesis, through ball milling of the neat reagents. To save experimental efforts, preliminary theoretical screening of the sensor-substrate ensemble can help in predicting host–guest interactions in silico and improve the design of the final MOF structure, as already observed in the field of CO_2 storage and utilization [84]. As for the metal type, the best candidate is surely zirconium, since the octahedral [Zr_6] metallic node present in all members of the "UiO family" generates MOF structures featured by high chemical and thermal stability. At present, our research group is proceeding in this direction for the obtainment of well-performing MOFs in the exciting field of luminescent sensing.

Author Contributions: Conceptualization, A.R.; literature data collection and curation, G.M., G.G. and A.R.; writing—original draft preparation, A.R.; writing—review and editing, G.M. and G.G.; funding acquisition, A.R. and G.G.

Funding: Our research on synthesis of thiazole-based MOFs and applications in gas storage and luminescent sensing is funded by the 2018–2021 bilateral CNR-RFBR project of the Italian National Research Council (CNR) and by the Italian MIUR through the PRIN 2017 Project Multi-e (20179337R7) "Multielectron transfer for the conversion of small molecules: an enabling technology for the chemical use of renewable energy". G.G. thanks the TRAINER project (Catalysts for Transition to Renewable Energy Future) Ref. ANR-17-MPGA-0017 for financial support.

Conflicts of Interest: The authors declare no conflict of interest.

References

1. Blasse, G.; Grabmaier, B. *Luminescent Materials*; Springer: Berlin, Germany, 1994.
2. Bünzli, J.-C.G. Lanthanide Luminescence for Biomedical Analyses and Imaging. *Chem. Rev.* **2010**, *110*, 2729–2755. [CrossRef] [PubMed]
3. You, L.; Zha, D.; Anslyn, E.V. Recent Advances in Supramolecular Analytical Chemistry Using Optical Sensing. *Chem. Rev.* **2015**, *115*, 7840–7892. [CrossRef] [PubMed]

4. Batten, S.R.; Champness, N.R.; Chen, X.-M.; Garcia-Martinez, J.; Kitagawa, S.; Öhrström, L.; O'Keeffe, M.; Suh, M.P.; Reedijk, J. Terminology of Metal–Organic Frameworks and Coordination Polymers (Iupac Recommendations 2013). *Pure Appl. Chem.* **2013**, *85*, 1715–1724. [CrossRef]
5. Cui, Y.; Yue, Y.; Qian, G.; Chen, B. Luminescent Functional Metal–Organic Frameworks. *Chem. Rev.* **2012**, *112*, 1126–1162. [CrossRef] [PubMed]
6. Ye, J.-W.; Zhou, H.-L.; Liu, S.-Y.; Cheng, X.-N.; Lin, R.-B.; Qi, X.-L.; Zhang, J.-P.; Chen, X.-M. Encapsulating Pyrene in a Metal–Organic Zeolite for Optical Sensing of Molecular Oxygen. *Chem. Mater.* **2015**, *27*, 8255–8260. [CrossRef]
7. Zhang, Y.; Yuan, S.; Day, G.; Wang, X.; Yang, X.; Zhou, H.-C. Luminescent Sensors Based on Metal–Organic Frameworks. *Coord. Chem. Rev.* **2018**, *354*, 28–45. [CrossRef]
8. Lustig, W.P.; Mukherjee, S.; Rudd, N.D.; Desai, A.V.; Li, J.; Ghosh, S.K. Metal–Organic Frameworks: Functional Luminescent and Photonic Materials for Sensing Applications. *Chem. Soc. Rev.* **2017**, *46*, 3242–3285. [CrossRef]
9. Yi, F.; Chen, D.; Wu, M.; Han, L.; Jiang, H.-L. Chemical Sensors Based on Metal–Organic Frameworks. *ChemPlusChem* **2016**, *81*, 675–690. [CrossRef]
10. Allendorf, M.D.; Bauer, C.A.; Bhakta, R.K.; Houk, R.J.T. Luminescent Metal–Organic Frameworks. *Chem. Rev.* **2009**, *38*, 1330–1352. [CrossRef]
11. Cepeda, J.; Rodríguez-Diéguez, A. Tuning the Luminescence Performance of Metal–Organic Frameworks Based on d^{10} Metal Ions: From an Inherent Versatile Behaviour to their Response to External Stimuli. *CrystEngComm* **2016**, *18*, 8556–8573. [CrossRef]
12. Barbieri, A.; Accorsi, G.; Armaroli, N. Luminescent Complexes Beyond the Platinum Group: The d^{10} Avenue. *Chem. Commun.* **2008**, 2185–2193. [CrossRef] [PubMed]
13. Cui, Y.; Chen, B.; Qian, G. Lanthanide Metal-Organic Frameworks for Luminescent Sensing and Light-Emitting Applications. *Coord. Chem. Rev.* **2014**, *273–274*, 76–86. [CrossRef]
14. San Sebastian, E.; Rodríguez-Diéguez, A.; Seco, J.M.; Cepeda, J. Coordination Polymers with Intriguing Photoluminescence Behavior: The Promising Avenue for Greatest Long-Lasting Phosphors. *Eur. J. Inorg. Chem.* **2018**, 2155–2174. [CrossRef]
15. Jiao, J.; Gong, W.; Wu, X.; Yang, S.; Cui, Y. Multivariate Crystalline Porous Materials: Synthesis, Property and Potential Application. *Coord. Chem. Rev.* **2019**, *385*, 174–190. [CrossRef]
16. Du, M.; Li, C.-P.; Liu, C.-S.; Fang, S.-M. Design and Construction of Coordination Polymers with Mixed-Ligand Synthetic Strategy. *Coord. Chem. Rev.* **2013**, *257*, 1282–1305. [CrossRef]
17. Lescouet, T.; Kockrick, E.; Bergeret, G.; Pera-Titus, M.; Aguado, S.; Farrusseng, D. Homogeneity of Flexible Metal–Organic Frameworks Containing Mixed Linkers. *J. Mater. Chem.* **2012**, *22*, 10287–10293. [CrossRef]
18. Travlou, N.L.A.; Singh, K.; Rodriguez-Castellon, E.; Bandosz, T.J. Cu–BTC MOF–Graphene-Based Hybrid Materials as Low Concentration Ammonia Sensors. *J. Mater. Chem. A* **2015**, *3*, 11417–11429. [CrossRef]
19. Desai, A.V.; Samanta, P.; Manna, B.; Ghoh, S.K. Aqueous Phase Nitric Oxide Detection by an Amine-Decorated Metal–Organic Framework. *Chem. Commun.* **2015**, *51*, 6111–6114. [CrossRef]
20. Gassensmith, J.J.; Kim, J.Y.; Holcroft, J.M.; Farha, O.K.; Stoddart, J.F.; Hupp, J.T.; Jeong, N.C. A Metal–Organic Framework-Based Material for Electrochemical Sensing of Carbon Dioxide. *J. Am. Chem. Soc.* **2014**, *136*, 8277–8282. [CrossRef]
21. Cui, J.; Wong, Y.-L.; Zeller, M.; Hunter, A.D.; Xu, Z. Pd Uptake and H_2S Sensing by an Amphoteric Metal–Organic Framework with a Soft Core and Rigid Side Arms. *Angew. Chem. Int. Ed.* **2014**, *53*, 14438–14442. [CrossRef]
22. Barrett, S.M.; Wang, C.; Lin, W. Oxygen Sensing via Phosphorescence Quenching of Doped Metal–Organic Frameworks. *J. Mater. Chem.* **2012**, *22*, 10329–10334. [CrossRef]
23. Xie, Z.; Ma, L.; deKrafft, K.E.; Jin, A.; Lin, W. Porous Phosphorescent Coordination Polymers for Oxygen Sensing. *J. Am. Chem. Soc.* **2010**, *132*, 922–923. [CrossRef] [PubMed]
24. Zhang, M.; Feng, G.; Song, Z.; Zhou, Y.P.; Chao, H.Y.; Yuan, D.; Tan, T.T.; Guo, Z.; Hu, Z.; Tang, B.Z.; et al. Two-Dimensional Metal–Organic Framework with Wide Channels and Responsive Turn-On Fluorescence for the Chemical Sensing of Volatile Organic Compounds. *J. Am. Chem. Soc.* **2014**, *136*, 7241–7244. [CrossRef] [PubMed]
25. Li, Y.; Zhang, S.; Song, D. A Luminescent Metal–Organic Framework as a Turn-On Sensor for DMF Vapor. *Angew. Chem. Int. Ed.* **2013**, *52*, 710–713. [CrossRef]

26. Shustova, N.B.; McCarthy, B.D.; Dinca, M. Metal–Organic Frameworks: An Alternative to Aggregation-Induced Emission. *J. Am. Chem. Soc.* **2011**, *133*, 20126–20129. [CrossRef] [PubMed]
27. Hong, Y.; Lam, J.W.Y.; Tang, B.Z. Aggregation-Induced Emission. *Chem. Soc. Rev.* **2011**, *40*, 5361–5388. [CrossRef]
28. Zhao, S.S.; Yang, J.; Liu, Y.Y.; Ma, J.F. Fluorescent Aromatic Tag-Functionalized MOFs for Highly Selective Sensing of Metal Ions and Small Organic Molecules. *Inorg. Chem.* **2016**, *55*, 2261–2273. [CrossRef]
29. Shustova, N.B.; Cozzolino, A.F.; Reineke, S.; Baldo, M.; Dinca, M. Selective Turn-On Ammonia Sensing Enabled by High-Temperature Fluorescence in Metal–Organic Frameworks with Open Metal Sites. *J. Am. Chem. Soc.* **2013**, *135*, 13326–13329. [CrossRef]
30. Mallick, A.; Garai, B.; Addicoat, M.A.; Petkov, P.S.; Heine, T.; Banerjee, R. Solid State Organic Amine Detection in a Photochromic Porous Metal Organic Framework. *Chem. Sci.* **2015**, *6*, 1420–1425. [CrossRef]
31. Takashima, Y.; Martinez, V.M.; Furukawa, S.; Kondo, M.; Shimomura, S.; Uehara, H.; Nakahama, M.; Sugimoto, K.; Kitagawa, S. Molecular Decoding Using Luminescence from an Entangled Porous Framework. *Nat. Commun.* **2011**, *2*, 168. [CrossRef]
32. Stylianou, K.C.; Heck, R.; Chong, S.Y.; James, J.B.; Jones, T.A.; Khimyak, Y.Z.; Bradshaw, D.; Rosseinsky, M.J. A Guest-Responsive Fluorescent 3D Microporous Metal–Organic Framework Derived from a Long-Lifetime Pyrene Core. *J. Am. Chem. Soc.* **2010**, *132*, 4119–4130. [CrossRef] [PubMed]
33. Mukherjee, S.; Desai, A.V.; Manna, B.; Inamdar, A.I.; Ghosh, S.K. Exploitation of Guest Accessible Aliphatic Amine Functionality of a Metal–Organic Framework for Selective Detection of 2,4,6-Trinitrophenol (TNP) in Water. *Cryst. Growth Des.* **2015**, *15*, 4627–4634. [CrossRef]
34. Nagarkar, S.S.; Desai, A.V.; Samanta, P.; Ghosh, S.K. Aqueous Phase Selective Detection of 2,4,6-Trinitrophenol Using a Fluorescent Metal–Organic Framework with a Pendant Recognition Site. *Dalton Trans.* **2015**, *44*, 15175–15180. [CrossRef] [PubMed]
35. Guo, Y.; Feng, X.; Han, T.; Wang, S.; Lin, Z.; Dong, Y.; Wang, B. Tuning the Luminescence of Metal–Organic Frameworks for Detection of Energetic Heterocyclic Compounds. *J. Am. Chem. Soc.* **2014**, *136*, 15485–15488. [CrossRef] [PubMed]
36. Qin, L.; Lin, L.-X.; Fang, Z.-P.; Yang, S.-P.; Qiu, G.-H.; Chen, J.-X.; Chen, W.-H. A Water-Stable Metal–Organic Framework of a Zwitterionic Carboxylate with Dysprosium: A Sensing Platform for Ebolavirus RNA Sequences. *Chem. Commun.* **2016**, *52*, 132–135. [CrossRef]
37. Zhang, H.T.; Zhang, J.W.; Huang, G.; Du, Z.Y.; Jiang, H.L. An Amine-Functionalized Metal–Organic Framework as a Sensing Platform for DNA Detection. *Chem. Commun.* **2014**, *50*, 12069–12072. [CrossRef]
38. Zhang, S.-R.; Li, J.; Du, D.-Y.; Qin, J.-S.; Li, S.-L.; He, W.-W.; Su, Z.-M.; Lan, Y.-Q. A Multifunctional Microporous Anionic Metal–Organic Framework for Column-Chromatographic Dye Separation and Selective Detection and Adsorption of Cr^{3+}. *J. Mater. Chem. A* **2015**, *3*, 23426–23434. [CrossRef]
39. Wang, Y.; Cheng, L.; Liu, Z.-Y.; Wang, X.-G.; Ding, B.; Yin, L.; Zhou, B.-B.; Li, M.-S.; Wang, J.-X.; Zhao, X.-J. An Ideal Detector Composed of Two-Dimensional Cd(II)-Triazole Frameworks for Nitro-Compound Explosives and Potassium Dichromate. *Chem. Eur. J.* **2015**, *21*, 14171–14178. [CrossRef]
40. Wu, P.; Liu, Y.; Liu, Y.; Wang, J.; Li, Y.; Liu, W.; Wang, J. Cadmium-Based Metal–Organic Framework as a Highly Selective and Sensitive Ratiometric Luminescent Sensor for Mercury(II). *Inorg. Chem.* **2015**, *54*, 11046–11048. [CrossRef]
41. Hao, J.-N.; Yan, B. A Water-Stable Lanthanide-Functionalized MOF as a Highly Selective and Sensitive Fluorescent Probe for Cd^{2+}. *Chem. Commun.* **2015**, *51*, 7737–7740. [CrossRef]
42. Cui, L.; Wu, J.; Li, J.; Ju, H. Electrochemical Sensor for Lead Cation Sensitized with a DNA Functionalized Porphyrinic Metal–Organic Framework. *Anal. Chem.* **2015**, *87*, 10635–10641. [CrossRef] [PubMed]
43. Wang, Y.; Wu, Y.; Xie, J.; Hu, X. Metal–Organic Framework Modified Carbon Paste Electrode for Lead Sensor. *Sens. Actuators B* **2013**, *177*, 1161–1166. [CrossRef]
44. Tan, H.; Liu, B.; Chen, Y. Lanthanide Coordination Polymer Nanoparticles for Sensing of Mercury(II) by Photoinduced Electron Transfer. *ACS Nano* **2012**, *6*, 10505–10511. [CrossRef] [PubMed]
45. Ma, J.-P.; Yu, Y.; Dong, Y.-B. Fluorene-Based Cu(II)-MOF: A Visual Colorimetric Anion Sensor and Separator Based on an Anion-Exchange Approach. *Chem. Commun.* **2012**, *48*, 2946–2948. [CrossRef] [PubMed]
46. Chow, C.-F.; Lam, M.H.W.; Wong, W.-Y. A Heterobimetallic Ruthenium(II)–Copper(II) Donor–Acceptor Complex as a Chemodosimetric Ensemble for Selective Cyanide Detection. *Inorg. Chem.* **2004**, *43*, 8387–8393. [CrossRef] [PubMed]

47. Robin, A.Y.; Fromm, K.M. Coordination Polymer Networks with O- And N-Donors: What They Are, Why and How They Are Made. *Coord. Chem. Rev.* **2006**, *250*, 2127–2157. [CrossRef]
48. Janiak, C. Engineering Coordination Polymers Towards Applications. *Dalton Trans.* **2003**, 2781–2804. [CrossRef]
49. Esteves, C.I.C.; Silva, A.M.F.; Raposo, M.M.M.; Costa, S.P.G. Unnatural Benz-X-Azolyl Asparagine Derivatives as Novel Fluorescent Amino Acids: Synthesis and Photophysical Characterization. *Tetrahedron* **2009**, *65*, 9373–9377. [CrossRef]
50. Batista, R.M.F.; Costa, S.P.G.; Raposo, M.M.M. Synthesis of New Fluorescent 2-(2′,2″-Bithienyl)-1,3-Benzothiazoles. *Tetrahedron Lett.* **2004**, *45*, 2825–2828. [CrossRef]
51. Lee, S.-H.; Otomo, A.; Nakahama, T.; Yamada, T.; Kamikado, T.; Yokoyama, S.; Mashiko, S. Novel Rigid-Rod 2,6-Diphenylbenzo[1,2-*d*:4,5-*d*′]Bisthiazole (DPBBT) Derivatives for Second-Order Nonlinear Optical Chromophores. *J. Mater. Chem.* **2002**, *12*, 2187–2188. [CrossRef]
52. Breitung, E.M.; Shu, C.-F.; McMahon, R.J. Thiazole and Thiophene Analogues of Donor−Acceptor Stilbenes: Molecular Hyperpolarizabilities and Structure−Property Relationships. *J. Am. Chem. Soc.* **2000**, *122*, 1154–1160. [CrossRef]
53. Sangiorgi, N.; Sangiorgi, A.; Dessì, A.; Zani, L.; Calamante, M.; Reginato, G.; Mordini, A.; Sanson, A. Improving the Efficiency Of Thin-Film Fiber-Shaped Dye-Sensitized Solar Cells By Using Organic Sensitizers. *Sol. Energy Mater. Sol. Cells* **2020**, *204*, 110209. [CrossRef]
54. Dessì, A.; Monai, M.; Bessi, M.; Montini, T.; Calamante, M.; Mordini, M.; Reginato, G.; Trono, C.; Fornasiero, P.; Zani, L. Towards Sustainable H_2 Production: Rational Design of Hydrophobic Triphenylamine-based Dyes for Sensitized Ethanol Photoreforming. *ChemSusChem* **2018**, *11*, 793–805. [CrossRef] [PubMed]
55. Staderini, S.; Tuci, G.; Luconi, L.; Müller, P.; Kaskel, S.; Eychmüller, A.; Eichler, F.; Giambastiani, G.; Rossin, A. Zinc Coordination Polymers Containing Isomeric Forms of *p*-(Thiazolyl)benzoic Acid: Blue-Emitting Materials with a Solvatochromic Response to Water. *Eur. J. Inorg. Chem.* **2017**, 4909–4918. [CrossRef]
56. Staderini, S.; Tuci, G.; D'Angelantonio, M.; Manoli, F.; Manet, I.; Giambastiani, G.; Peruzzini, M.; Rossin, A. Zinc Coordination Polymers Containing the *m*-(2-thiazolyl)benzoic Acid Spacer: Synthesis, Characterization and Luminescent Properties in Aqueous Solutions. *ChemistrySelect* **2016**, *6*, 1123–1131. [CrossRef]
57. Tuci, G.; Giambastiani, G.; Kwon, S.; Stair, P.C.; Snurr, R.Q.; Rossin, A. Chiral Co(II) Metal−Organic Framework in the Heterogeneous Catalytic Oxidation of Alkenes under Aerobic and Anaerobic Conditions. *ACS Catal.* **2014**, *4*, 1032–1039. [CrossRef]
58. Rossin, A.; Tuci, G.; Giambastiani, G.; Peruzzini, M. 1D and 2D Thiazole-Based Copper(II) Coordination Polymers: Synthesis and Applications in Carbon Dioxide Capture. *ChemPlusChem* **2014**, *79*, 406–412. [CrossRef]
59. Rossin, A.; Di Credico, B.; Giambastiani, G.; Peruzzini, A.; Pescitelli, G.; Reginato, G.; Borfecchia, E.; Gianolio, D.; Lamberti, C.; Bordiga, S. Synthesis, Characterization and CO_2 Uptake of a Chiral Co(II) Metal–Organic Framework Containing a Thiazolidine-Based Spacer. *J. Mater. Chem.* **2012**, *22*, 10335–10344. [CrossRef]
60. Rossin, A.; Giambastiani, G. Structural Features and Applications of Metal–Organic Frameworks Containing Thiazole- and Thiazolidine-Based Spacers. *CrystEngComm* **2015**, *17*, 218–228. [CrossRef]
61. Dannenbauer, N.; Matthes, P.R.; Scheller, T.P.; Nitsch, J.; Zottnick, S.H.; Gernert, M.S.; Steffen, A.; Lambert, C.; Müller-Buschbaum, K. Near-Infrared Luminescence and Inner Filter Effects of Lanthanide Coordination Polymers with 1,2-Di(4-pyridyl)Ethylene. *Inorg. Chem.* **2015**, *55*, 7396–7406. [CrossRef]
62. Dannenbauer, N.; Matthes, P.R.; Müller-Buschbaum, K. Luminescent Coordination Polymers for the VIS and NIR Range Constituting $LnCl_3$ and 1,2-Bis(4-Pyridyl)Ethane. *Dalton Trans.* **2016**, *45*, 6529–6540. [CrossRef] [PubMed]
63. Dannenbauer, N.; Zottnick, S.H.; Müller-Buschbaum, K. Thiazole and the Diazines Pyrazine and Pyrimidine as Sensitizers for Lanthanide Luminescence from VIS to NIR. *Z. Anorg. Allg. Chem.* **2017**, *643*, 1513–1518. [CrossRef]
64. Chen, M.; Sun, R.; Ye, Y.; Tang, H.; Dong, X.; Yan, J.; Wang, K.; Zhou, Q.; Wang, Z. Application of a Novel Red-Emitting Cationic Iridium(III) Coordination Polymer in Warm White Light-Emitting Diodes. *Opt. Mater.* **2018**, *76*, 141–146. [CrossRef]
65. Artem'ev, A.V.; Samsonenko, D.G.; Antonova, O.V. CuI-Based Coordination Polymers with 2-Thiazolyl Sulfide Ligands: First Examples. *Polyhedron* **2018**, *151*, 171–176. [CrossRef]

66. Rogovoy, M.I.; Samsonenko, D.G.; Rakhmanova, M.I.; Artem'ev, A.V. Self-Assembly of Ag(I)-Based Complexes and Layered Coordination Polymers Bridged by (2-Thiazolyl)Sulfides. *Inorg. Chim. Acta* **2019**, *489*, 19–26. [CrossRef]
67. Woodward, A.N.; Kolesar, J.M.; Hall, S.R.; Saleh, N.; Jones, D.S.; Walter, M.G. Thiazolothiazole Fluorophores Exhibiting Strong Fluorescence and Viologen-Like Reversible Electrochromism. *J. Am. Chem. Soc.* **2017**, *139*, 8467–8473. [CrossRef]
68. Zhang, Z.; Chen, Y.; Hung, W.; Tang, W.; Hsu, Y.; Chen, C.; Meng, F.; Chou, P. Control of the Reversibility of Excited-State Intramolecular Proton Transfer (ESIPT) Reaction: Host-Polarity Tuning White Organic Light Emitting Diode on a New Thiazolo[5,4-d]thiazole ESIPT System. *Chem. Mater.* **2016**, *23*, 8815–8824. [CrossRef]
69. Dessì, A.; Calamante, M.; Mordini, A.; Peruzzini, M.; Sinicropi, A.; Basosi, R.; Fabrizi de Biani, F.; Taddei, M.; Colonna, D.; di Carlo, A.; et al. Organic Dyes with Intense Light Absorption Especially Suitable for Application in Thin-Layer Dye-Sensitized Solar Cells. *Chem. Commun.* **2014**, *50*, 13952–13955. [CrossRef]
70. Pereira Clementino, R.F.; de Souza Santos, A.B.; Bandeira Jovino Marques, O.J.; Ratkovski, D.R.; Gatto, C.C.; Malvestiti, I.; de Araujo Machado, F.L.; Lago Falcão, E.H. Structural Description, Luminescent and Magnetic Properties of Novel 2-D Coordination Polymers Containing Thiazolo[5,4-*d*]Thiazole Rings and Trivalent Lanthanide Ions. *J. Solid State Chem.* **2018**, *268*, 94–101. [CrossRef]
71. Zhai, Z.; Yang, S.; Cao, M.; Li, L.; Du, C.; Zang, S. Rational Design of Three Two-Fold Interpenetrated Metal–Organic Frameworks: Luminescent Zn/Cd-Metal–Organic Frameworks for Detection of 2,4,6-Trinitrophenol and Nitrofurazone in the Aqueous Phase. *Cryst. Growth Des.* **2018**, *18*, 7173–7182. [CrossRef]
72. Zhai, Z.; Yang, S.; Luo, P.; Li, L.; Du, C.; Zang, S. Dicarboxylate-Induced Structural Diversity of Luminescent Zn(II)/Cd(II) Metal–Organic Frameworks Based on the 2,5-Bis(4-Pyridyl)Thiazolo[5,4-d]Thiazole Ligand. *Eur. J. Inorg. Chem.* **2019**, 2725–2734. [CrossRef]
73. Khatun, A.; Panda, D.K.; Sayresmith, N.; Walter, M.G.; Saha, S. Thiazolothiazole-Based Luminescent Metal–Organic Frameworks with Ligand-to-Ligand Energy Transfer and Hg^{2+} Sensing Capabilities. *Inorg. Chem.* **2019**, *58*, 12707–12715. [CrossRef] [PubMed]
74. Arici, M.; Yesilel, O.Z.; Dege, N. Three Co(II) Coordination Polymers Constructed from 2,5-Di(4-Pyridyl)Thiazolo[5,4-*d*]Thiazole and V-Shaped Dicarboxylic Acids: Syntheses, Characterizations, Structural Diversity and Optical Properties. *Polyhedron* **2019**, *163*, 77–83. [CrossRef]
75. Li, P.; Yin, X.; Gao, L.; Yang, S.; Sui, Q.; Gong, T.; Gao, E. Modulating Excitation Energy of Luminescent Metal–Organic Frameworks for Detection of Cr(VI) in Water. *ACS Appl. Nano Mater.* **2019**, *2*, 4646–4654. [CrossRef]
76. Fan, J.; Li, J.; Zhang, L.; Zhang, L.; Wang, D. New Cu^{II} and Cd^{II} Metal–organic Coordination Polymers with 1,2,4-Triazolo[3,4-b]-1,3,4-Thiadiazole Ligands: Syntheses, Structures and Luminescent Properties. *J. Chin. Chem. Soc.* **2015**, *62*, 786–792. [CrossRef]
77. Wen, G.; Liu, D.; Chen, Y.; Wei, Y.; Zhu, Q.; Wang, X.; Xua, M.; Yao, Y.; Ma, L. A Rare Twofold Interpenetrating NbO Mixed-Ligand Mesomeric Network from Two Individual Heterochiral 3D Frameworks. *Inorg. Chem. Commun.* **2016**, *74*, 86–89. [CrossRef]
78. Zhang, X.; Zhang, X.; Fan, C.; Zong, Z.; Zhang, D.; Luo, Q.; Bi, C.; Fan, Y. A Novel Metal–Organic Frameworks Assembled by One Angular Ligand and 5-Aminoisophthalic Acid: Synthesis, Structure, Electrochemical and Photocatalytic Properties. *Polyhedron* **2019**, *168*, 21–27. [CrossRef]
79. Sk, M.; Biswas, S. A Thiadiazole-Functionalized Zr(IV)-Based Metal–Organic Framework as a Highly Fluorescent Probe for the Selective Detection of Picric Acid. *CrystEngComm* **2016**, *18*, 3104–3113. [CrossRef]
80. Mallick, A.; El-Zohry, A.M.; Shekhah, O.; Yin, J.; Jia, J.; Himanshu Aggarwal, H.; Emwas, A.; Mohammed, O.F.; Eddaoudi, M. Unprecedented Ultralow Detection Limit of Amines using a Thiadiazole-Functionalized Zr(IV)-Based Metal–Organic Framework. *J. Am. Chem. Soc.* **2019**, *141*, 7245–7249. [CrossRef]
81. Song, W.; Liang, L.; Cui, X.; Wang, X.; Yang, E.; Zhao, X. Assembly of Zn^{II}-Coordination Polymers Constructed from Benzothiadiazole Functionalized Bipyridines and V-Shaped Dicarboxylic Acids: Topology Variety, Photochemical and Visible-Light-Driven Photocatalytic Properties. *CrystEngComm* **2018**, *20*, 668–678. [CrossRef]

82. Wang, X.; Xiong, Y.; Sha, X.; Liu, G.; Lin, H. Various Polycarboxylate-Directed Cd(II) Coordination Polymers Based on a Semirigid Bis-Pyridyl-bis-Amide Ligand: Construction and Fluorescent and Photocatalytic Properties. *Cryst. Growth Des.* **2017**, *17*, 483–496. [CrossRef]
83. Gogia, A.; Mandal, S.K. A Rational Design and Green Synthesis of 3D Metal Organic Frameworks Containing a Rigid Heterocyclic Nitrogen-Rich Dicarboxylate: Structural Diversity, CO_2 Sorption and Selective Sensing of 2,4,6-TNP in Water. *Dalton Trans.* **2019**, *48*, 2388–2398. [CrossRef] [PubMed]
84. Müller, P.; Bucior, B.; Tuci, G.; Luconi, L.; Getzschmann, J.; Kaskel, S.; Snurr, R.Q.; Giambastiani, G.; Rossin, A. Computational Screening, Synthesis and Testing of Metal–Organic Frameworks with a Bithiazole Linker for Carbon Dioxide Capture and its Green Conversion into Cyclic Carbonates. *Mol. Syst. Des. Eng.* **2019**, *4*, 1000–1013. [CrossRef]

inorganics MDPI

Review

Metal–Organic Frameworks as Versatile Platforms for Organometallic Chemistry

Fan Chen [1], Hannah F. Drake [1], Liang Feng [1], Joshua A. Powell [1], Kun-Yu Wang [1], Tian-Hao Yan [1] and Hong-Cai Zhou [1,2,*]

1 Department of Chemistry, Texas A&M University, College Station, TX 77843-3255, USA; fanchen2018@tamu.edu (F.C.); hfd100@tamu.edu (H.F.D.); fengliang@tamu.edu (L.F.); jpowell7@tamu.edu (J.A.P.); wangkunyu@tamu.edu (K.-Y.W.); thyan426@tamu.edu (T.-H.Y.)
2 Department of Materials Science and Engineering, Texas A&M University, College Station, TX 77843-3003, USA
* Correspondence: zhou@chem.tamu.edu

Abstract: Metal–organic frameworks (MOFs) are emerging porous materials with highly tunable structures developed in the 1990s, while organometallic chemistry is of fundamental importance for catalytic transformation in the academic and industrial world for many decades. Through the years, organometallic chemistry has been incorporated into functional MOF construction for diverse applications. Here, we will focus on how organometallic chemistry is applied in MOF design and modifications from linker-centric and metal-cluster-centric perspectives, respectively. Through structural design, MOFs can function as a tailorable platform for traditional organometallic transformations, including reaction of alkenes, cross-coupling reactions, and C–H activations. Besides, an overview will be made on other application categories of organometallic MOFs, such as gas adsorption, magnetism, quantum computing, and therapeutics.

Keywords: metal–organic framework; organometallic; coordination bonds; supramolecular chemistry; catalysis

Citation: Chen, F.; Drake, H.F.; Feng, L.; Powell, J.A.; Wang, K.-Y.; Yan, T.-H.; Zhou, H.-C. Metal–Organic Frameworks as Versatile Platforms for Organometallic Chemistry. *Inorganics* **2021**, *9*, 27. https://doi.org/10.3390/inorganics9040027

Academic Editor: Andrea Rossin

Received: 2 March 2021
Accepted: 1 April 2021
Published: 9 April 2021

1. Introduction

Catalytic transformations to design new and tailored commodity items have been of interest for many decades, both in the private, industrial, and academic world. This success is owed in part to the catalytic transformations of organometallic complexes, their functionalizable environments, and their active centers. Many important industrial processes that we rely upon in our modern lifestyle depend on the basic chemical transformations using organometallic chemical processes. In particular, the chemical transformations used to generate ammonia, synthesis gas, epoxyethane, sulfuric acid, nitric acid, and many polymer-based plastics involve an organometallic catalyst (iron, nickel, silver on alumina, vanadium, platinum and rhodium, and titanium respectively) [1]. It is estimated that 300 million tons of polymer-based plastics alone are produced globally each year at least in part from catalytic synthesis processes [2–4]. This area of research has long been a key feature in the modernization and globalization of our world as many of the everyday commodities we utilize are made through organometallic chemical processes. Thus, there are also many new approaches to design better and more effective ways of producing these commodities by further refining the organometallic catalytic processes that are utilized most heavily today. Among these approaches, Metal–organic frameworks (MOFs) have taken a central role in the improvement of organometallic catalytic research.

Within our context, MOFs, also known as porous coordination networks (PCNs), are a type of highly crystalline, well ordered coordination network compounds derived from both metal and organic components. MOFs originated in the late 1990s with seminal works published from Yaghi and Kitagawa and is thus still a very recent advance towards catalytic chemical systems [5,6]. Despite their relative infancy as a field, these materials are rapidly becoming well known for their versatility due to their diverse tunability of the linkers, pore voids, and metal nodes that make up the three key features of these structures [7–10]. Although MOFs are not considered to be traditional organometallic compounds in the most restrictive definition of organometallic chemistry, as many of them do not contain M–C bonds, the unique tunability and ability to design tailored chemical reactions allows MOFs to undergo many of the same chemical processes used in M–C type organometallic chemistry. Within MOF structures, most of the organic linker coordination bonds that hold the structures together are comprised of N, O, or S donor atom to metal cluster bonds [11]. Ultimately, many MOFs discussed in this review do perform catalytic organometallic M–C type chemistry as the scaffold material even though the scaffold itself is a coordination type complex [12–16].

The three main components of MOFs are their linkers, void spaces, and metal clusters. All three of these features can be tuned specifically towards desired transformations. For the purposes of this review, we will emphasize two main features of interest: the functionalization of organic linkers towards catalytic chemistry and the transformation of metal cluster centers towards these processes. We will expand upon the utilization of MOFs for traditional organometallic chemical applications, demonstrate the current directions of non-traditional utilization of MOFs in organometallic catalytic process, and give an outlook on the future directions of this field (Figure 1).

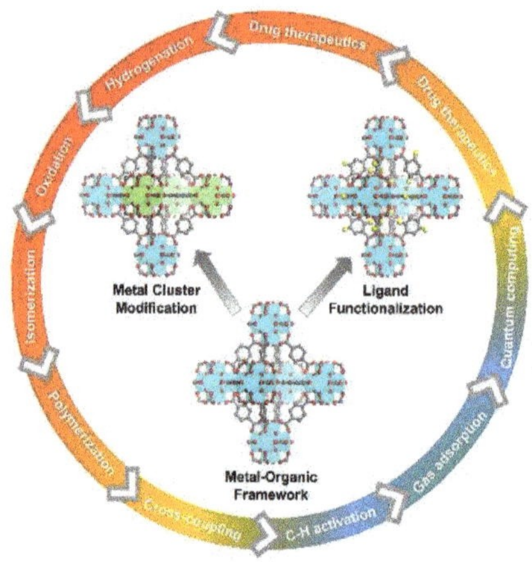

Figure 1. Summarizing figure of organometallic chemistry in metal–organic frameworks. Blue = Zr atoms. Green = Hf atoms. Red = O atoms. Grey = C atoms. Yellow = F atoms.

2. Linker-Based Functional Organometallic MOFs

MOFs are a class of crystalline porous material that have attracted considerable attention over the past few decades [17]. Due to their permanent porosity, large surface area, structural and functional diversity, and tunability, MOFs have been utilized in a diverse array of applications. Some of these applications include: gas sorption and separation, chemical sensing, proton conduction, and drug delivery [18–23]. Of particular interest, MOFs are considered to be promising candidates as heterogeneous organometallic catalysts [19–24]. For catalytic applications, MOFs offer two approaches: cluster mediated catalysis and linker mediated catalysis. The organic linker building block plays an important part in the reasonable design of MOF-based catalysts [25]. In this section, linker-based organometallic catalysis will be discussed. Additionally, the relationship between MOF structure and application will also be discussed.

2.1. Porphyrin-Based Linkers

Porphyrin chemistry has been of significant biological interest. It is no wonder than that porphyrin-based ligands would be of significant interest for MOF chemistry, particularly in light-harvesting, sensing, and catalysis [26]. As a result, porphyrin ligand-based MOFs have been substantially explored. Building porphyrins into MOF scaffolds eliminates the dimerization pathways and mitigates the solubility and chemical stability issues often associated with porphyrin chemistry. In return, porphyrins add additional functionality to the MOF scaffold, particularly in photo-sensing catalysis [27]. As a result, the introduction of porphyrins into MOFs as linkers has allowed for scientists to take advantage of both systems.

One example of a porphyrin-based MOFs came from the Zhou group, PCN-224. This MOF was assembled from 6-connected Zr_6 secondary building units (SBUs) and metalloporphyrin ligands (Fe-TCPP (TCPP = tetrakis(4-carboxyphenyl)-porphyrin)) (Figure 2), into a 3D nanochanneled, highly stable framework. PCN-224 has been utilized as a highly active, recoverable heterogeneous catalyst in CO_2/propylene oxide coupling reactions [28]. PCN-224(Ni) was tested to be extra stable for the remained property after 24 h immersion in over a wide range of pH in aqueous solution, from 1 to 11. Expanding upon this initial study, a series of PCN-224 analogues with ethyl, bromo, chloro, and fluoro to the β-position of porphyrin ligands was synthesized. This series was shown to be able to tune the chemical environment of the catalytic center for desired reactions by changing the electronic properties in the ligand scaffold. In this work it was also demonstrated that a substitution within the TCPP linker was able to increase the activity and selectivity of the catalytic iron sites in the 3-methylpentane oxidation catalysis [29].

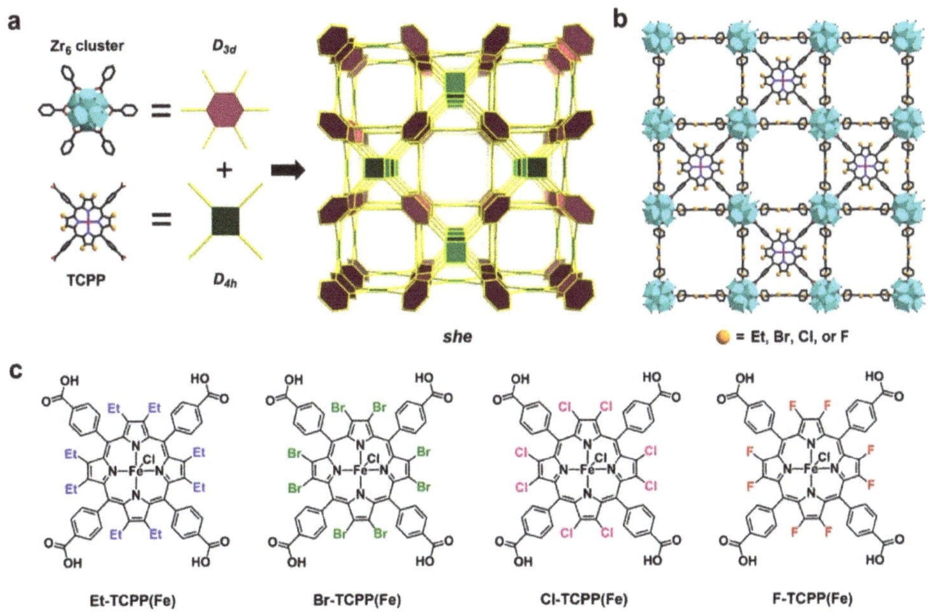

Figure 2. (**a**) Assembly of D_{3d} and D_{4h} nodes into (4,6)-connected *she* topological network. The Zr_6 cluster and TCPP(Fe) ligand can be viewed as 6- and 4-connected nodes, respectively. (**b**) View down the tetragonal channels of R-PCN-224(Fe) (black, carbon; purple, iron; blue, nitrogen; hydrogen and chloride, omitted). The yellow spheres represent Et, Br, Cl, or F. (**c**) Chemical structures of Et-TCPP(Fe), Br-TCPP(Fe), Cl-TCPP(Fe), and F-TCPP(Fe) linkers. Adapted with permission from [29]; Published by American Chemical Society, 2017.

Porphyrinic linkers can also be utilized as photosensitive building blocks in MOFs. Post-synthetic metalation of metal-free porphyrins is a common way of adjusting the catalytic property of porphyrins-based MOF. One particular example from the Rosseinsky group reported an aluminum-based porous porphyrin MOF that could perform water splitting under visible light irradiation. Within the rigid host structure of this MOF, the metal-free porphyrins could be metalated with Zn^{2+}, allowing for the catalytic generation of hydrogen and oxygen from water [30]. Another example of a post-synthetic metalation of porphyrin for tuned catalytic reactions is PCN-223, showing high stability in aqueous solutions with a wide range of pH due to its extremely high connectivity. This metal-free porphyrin Zr-MOF was metalated with Fe^{3+}, making it an excellent recyclable catalyst for hetero-Diels–Alder reactions [31].

In another work, a robust metalloporphyrinic MOF with perfluorophenylene functional groups, PCN-624, was synthesized utilizing rational structure design techniques. PCN-624 was constructed from 12-connected $[Ni_8(OH)_4(H_2O)_2Pz_{12}]$ (Pz = pyrazolide) nodes and fluorinated 5,10,15,20-tetrakis-(2,3,5,6-tetrafluoro-4-(1H-pyrazol-4-yl)phenyl)-porphyrin (TTFPPP) ligands. Pendant perfluorophenylene groups were fabricated onto the pore surface of PCN-624, resulting in the cyclable, selective synthesis of fullerene-anthracene bis-adducts. This example also demonstrates that MOFs can be utilized as a competitive platform for catalytic applications [32].

2.2. Polypyridyl-Based Linkers

In recent years, harvesting solar power through light harvesting materials has received interest in the scientific community in the application of renewable energy sources. The utilization of polypyridyl organometallic complexes in light harvesting devices is fairly well developed. As a result, MOFs designed for light harvesting have begun to incorporate polypyridyl moieties into their frameworks. This approach has attracted interest for applications in catalytic water oxidation, CO_2 reduction, photocatalytic hydrogen evolution, and the photocatalysis of organic reactions [25].

One of the pioneer research groups for doping polypyridyl organometallic complexes into MOFs, the Lin group, demonstrated the usability of this system for solar energy harvesting. In 2011, this group incorporated Ir, Re, and Ru polypyridyl complexes into the UiO-67 (UiO = University of Oslo) topology, leading to a series of highly active heterogenous catalysts for visible light-driven organic molecule transformations [33]. Later, the same group merged a Ir^{III} photoredox catalyst and a Ni^{II} cross-coupling catalyst into the same MOF scaffold to efficiently catalyze C-S bond formation between different aryl iodides and thiols [34].

Among organic transformations through organometallic catalysis, CO_2 reduction is one that is of particular interest, since CO_2 and CH_4 are widely known to be of use in chemical energy conversions. One approach to accomplish CO_2 reduction is through molecular iridium catalysts immobilized within a MOF. In one such study, Ir–UiO MOFs, mbpyOH–$IrCl_3$–UiO and mbpy$IrCl_3$–UiO (bpy = 2,2′-bipyridine), were synthesized and placed in a condensation chamber, facilitating CO_2 reduction [12]. In another study, a metal-organic layer (MOL), a freestanding monolayer of a 2D MOF, was utilized as a novel platform for CO_2 hydrogenation. As a result of modifications made to the Hf_{12} SBUs and the $[Ru(bpy)_3]^{2+}$ based linkers in the study, a new kind of MOL was prepared that showed efficient catalytic performance for the reduction of CO_2 [35] The structural stability of the MOLs was approved by the repeatable PXRD (powder X-ray diffraction) patterns after 24 h of CO_2 reduction.

An azide-functionalized UiO-66 was employed as a platform for the immobilization of a series of bidentate ligands on a MOF surface via an azide-alkyne "click" reaction. These immobilized ligands formed a highly effective mixed-ligand nickel catalyst. The product generated was demonstrated to be an effective heterogeneous organometallic catalyst for the Suzuki-Miyaura coupling reaction under mild conditions. The system was also shown to be recyclable with little to no change to the performance or properties of the original material [36].

An example of a metalloporphyrin MOF that is active towards C–H activation is PCN-602, which was reported by the Zhou group. This MOF is a base-resistant porphyrin-based system built from $[Ni_8(OH)_4(H_2O)_2Pz_{12}]$ (Pz = pyrazolate) clusters and 5,10,15,20-tetrakis(4-(pyrazolate-4-yl)-phenyl)porphyrin linkers [37]. The derived PCN-602(Mn) with Mn^{3+}-porphyrin centers was used to catalyze C–H bond chlorination and bromination reactions for cyclohexane and cyclopentane reactions, demonstrating good performance. The metalated porphyrin species as a catalyst was able to yield a 95% conversion rates under room temperature conditions after 2 h. These results indicated that this particular MOF was a highly effective catalyst for C–H halogenation of hydrocarbons under basic conditions. Moreover, PCN-602 exhibited great stability in aqueous solutions of OH^-, F^-, CO_3^{2-}, and PO_4^{3-}.

2.3. Pyridyl-Based Linkers

One of the earliest examples of a pyridyl-based MOFs comes from the Humphrey group who prepared the first organoarsine MOF, ACM-1, using a new pyridyl-functionalized triarylarsine ligand coordinated to Ni^{II} nodes. Under facile conditions, postsynthetic metalation of the ACM-1 structure was accomplished through the installation of dimeric Au_2Cl_2 complexes via the formation of As–Au bonds. Due to the rigidity of the MOF, the Au^I dimers displayed particularly short aurophilic bonds (2.76 Å) [38].

With growing focus on pyridyl-based MOFs, postsynthetic modifications are now fairly common. An example of this was with a Zr-MOF scaffold that was modified to obtain a bimetallic MOF with $MX_2(INA)_4$ moieties (INA = isonicotinate; M = Co^{2+} or Fe^{2+}; X = OH^-, Cl^-, Br^-, I^-, NCS^-, or $NCSe^-$). The step-by-step modification process not only changed the composition, symmetry, and unit cell of the MOF by introducing $MX_2(py)_4$ (py = pyridine), but the exchange also endowed interesting magnetic and electronic properties to the diamagnetic framework [39].

MOFs serve as designable platforms to construct complex coordination architectures for desirable applications. A method to place trans-coordinate metal centers with exposed equatorial positions in a MOF matrix has been reported. PCN-160, a Zr-based MOF, was initially synthesized and subsequently underwent postsynthetic modification in the form of ligand elimination and installation of pyridinecarboxylate ligands (Figure 3). The proximity between a pair of neighboring pyridyl groups within the ligand was suitable for the formation of trans-metal-binding sites that were capable of accepting multiple diatomic metal cations. Some metals tested in this body of work included: Ni^{2+}, Cu^{2+}, and Pd^{2+}. This particular trans-coordinated metal site in the MOF was utilized for the catalytic dimerization of ethylene. This catalyst in particular had high activity due to the exposed equatorial positions of the open metal sites [40].

2.4. Pincer-Based Linkers

Ligands that chelate a metal using three coplanar chelators are referred to as pincer ligands. In organometallic chemistry, pincer complexes have been extensively shown to be remarkable catalysts for a variety of applications [41]. It is no surprise then that pincer-based ligands have begun to be incorporated into MOFs in recent years. One such example is for a Zr-based MOF with modified Pd aryl diphosphinite (POCOP) pincer linkers. This particular system was shown to be remarkably active and recyclable as a catalyst for the transfer hydrogenation of benzaldeydes, together with the exceptional chemical stability even under strong acid (1 M HNO_3) and base (0.1 M NaOH) for weeks [15]. Typically pincer-based ligands require postsynthetic metalation of the chelation site otherwise the as synthesized MOF fails to form. There are very few reports of pincer-based MOFs that do not require this post synthetic step. One such MOF, a 3D MOF Co^{II} MOF was synthesized through first cyclometallation of the pincer ligand with $Pd^{II}Cl$ and subsequent use of the metalloligand as a starting material in a one pot solvothermal reaction. The Pd–Cl groups in the chelation site were highly catalytically active for CO_2 insertion at 1atm and 298 K [42].

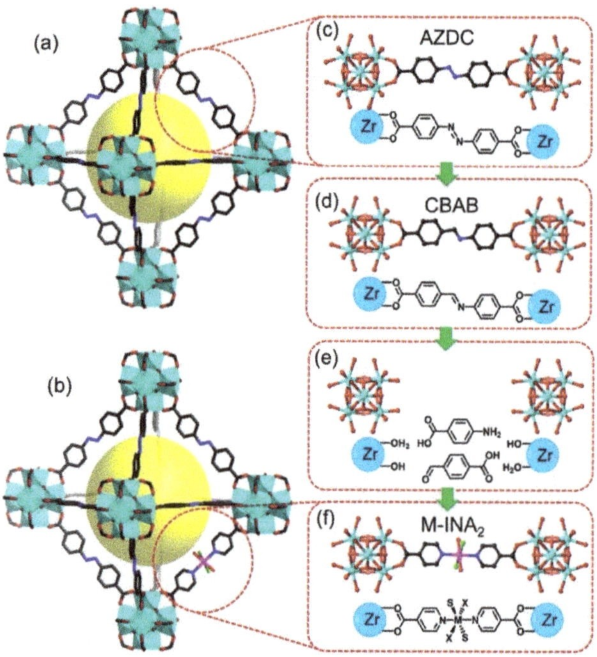

Figure 3. Structure of PCN-160 (**a**) and PCN-160-R%M with *trans*-chelating ligands (**b**). Transformation of ligand fragment in PCN-160 (**c**) by CBAB exchange (**d**), linker labialization (**e**), and installation of M-INA$_2$ (INA = isonicotinate) (**f**). These figures are based on respective single-crystal structures after removal of the disordered fragments. Cyan = Zr. Red = O. Grey = C. Blue = N. Adapted with permission from [40]; Published by American Chemical Society, 2018.

2.5. Other Linkers in MOFs

In addition to the porphyrins, polypyridyl, and pincer ligands discussed in this section, there have been other organometallic catalytic linkers that have been incorporated into MOFs. For example, N-heterocyclic carbene-based linkers (NHC) have been incorporated into MOF frameworks for catalytic applications. For these linkers, tuning the imidazole sidechains with different functional groups offers a dynamic array of possible catalytic activity. One example of an NHC-based MOF was published by the Zou group and contained an Iridium NHC metallolinker in a Zr-MOF. Through both direct synthesis and postsynthetic exchange, the MOF was able to be synthesized and adapted for allylic alcohol isomerization. This particular MOF demonstrated good catalytic activity and recyclability [43].

3. Cluster-Based Functional Organometallic MOFs

Metal clusters are widely known to be Lewis acidic sites within MOF structures. However, recent work has also explored the use of metal clusters within MOFs as organometallic catalysts and as a platform for the support of single-site organometallic catalysts. Single-site catalysts are preferred over metal oxide or metal nanoparticle catalysts for their greater catalytic selectivity. The most common method for creating single-site catalysts involves anchoring the catalytically active moiety onto a surface. This approach unfortunately tends to create problems with uniformity and surface-catalyst interactions [44]. MOFs have been presented as a solution to these problems as they are crystalline (i.e., highly uniform) materials and can be tuned with relative ease. Constraining catalytically active sites within MOFs can also prevent catalyst deactivation by ligand disproportionation. This type of functionalization leads to the formation of mixed-metal organic frameworks (M-MOFs). In this section, three methods for the functionalization of MOF clusters will be highlighted in regard to organometallic catalytic performance.

3.1. Post-Synthetic Metalation

Metals or organometallic species can be post-synthetically grafted onto a metal cluster. This technique is utilized when substitution within the cluster is not desired, but imparted functionality to the clusters is desired. Most often this technique is used to maintain structural stability in the framework where post-synthetic metal exchange (PSME) would result in structural instability or collapse. Post-synthetic metalation does not substitute metal atoms that already exist within the metal cluster, instead this technique grafts organometallic moieties onto the existing cluster in a structure. This approach can be accomplished through using techniques such as atomic layer deposition in MOFs (AIM) or through traditional organometallic chemical techniques. When considering the behavior of the organometallic moiety, the metal clusters in the scaffold can be treated as a both a ligand and as a support.

3.1.1. Platforms Suitable for Post-Synthetic Metalation

Group IV clusters, primarily $[Zr_6(\mu_3\text{-}O)_4(\mu_3\text{-}OH)_4(COO)_{12}]$ clusters (hereafter Zr_6 clusters), are the most common clusters used as supports for organometallic moieties. These clusters are common in MOFs, as they allow for high connectivity and are generally stable [45]. One of the earliest MOFs reported that contained a Group IV cluster was UiO-66. This structure contained 12-connected Zr_6 cluster joined by linear benzene dicarboxylate (bdc) linkers [46]. UiO-66 also has stable isostructural frameworks, UiO-67 and -68, which can be generated through isoreticular expansion of the organic ligands [46]. These MOFs have been extensively studied in the literature [46–50]. The Zr_6 cluster that is used in the UiO series has become perhaps the most popularly used Group IV cluster in MOF chemistry. A large number of Zr-MOFs: the PCN-700 series [51], NU-1000 (NU = Northwestern University) [52], MOF-545 [53], and NU-1200 [54], contain this cluster.

Zr-based clusters have several key advantages over other common clusters such as basic zinc carboxylates (BZCs) or paddlewheel clusters. Zr_6 clusters are tolerant to defects and can have high connectivity (up to 12 linkers) [45] compared to BZCs (up to 6 linkers) [55] or paddlewheel clusters (up to 4 linkers) [56]. This higher connectivity creates a greater degree of tunability for this structure as compared to lower connected clusters. In addition to the high connectivity, Zr_6 MOFs are also particularly stable due to their high cluster charge [45]. Although high connectivity is permitted by Zr_6 clusters, their tolerance to defects means that MOF variants with lower connectivity can also be designed without major loss to structural stability, which can lead to the presence of open coordination sites around the cluster that are available for post-synthetic metalation (Figure 4). In addition to the catalytic behavior and further functionalizability of the open coordination sites [57,58], coordinatively unsaturated metal clusters are also more likely to form flexible MOFs. This category of MOF is currently a hotbed for the study of selective or switchable catalysis [51,59]. Combining these properties with post-synthetic metalation of clusters creates exciting opportunities for the expansion of the range of reactions that can be catalyzed using Zr-MOFs.

3.1.2. Post-Synthetic Metalation Techniques

One commonly used technique for post-synthetic metalation is the use of AIM [60]. A famous example of the use of AIM in MOFs comes from the Hupp and Farha groups. In their work, they demonstrated the usability of NiS as a grafting material onto Zr_6 metal nodes in NU-1000 (Figure 5). In this method, nickel and sulfur precursors are vaporized and are deposited onto the MOF framework. The nickel binds to the hydroxyl/aqua ligand pairs in the first deposition step, and sulfur binds to the nickel in the second step [60]. Interestingly, the nickel species deposited by AIM are not distributed evenly across the structure. Clusters of nickel in the NU-1000 modified material were shown to gather predominantly within the smaller pores of the NU-1000 structure [61]. Following the deposition of nickel sulfide, the material could be used for photocatalytic water splitting experiments. It was found that the parent NU-1000 framework could not perform water splitting, but the NiS@NU-1000 composite material was highly active, evolving a significant amount of H_2 [60]. In addition to Ni, other metals have also been deposited using AIM, including Mo, Nb, and Co [62–64].

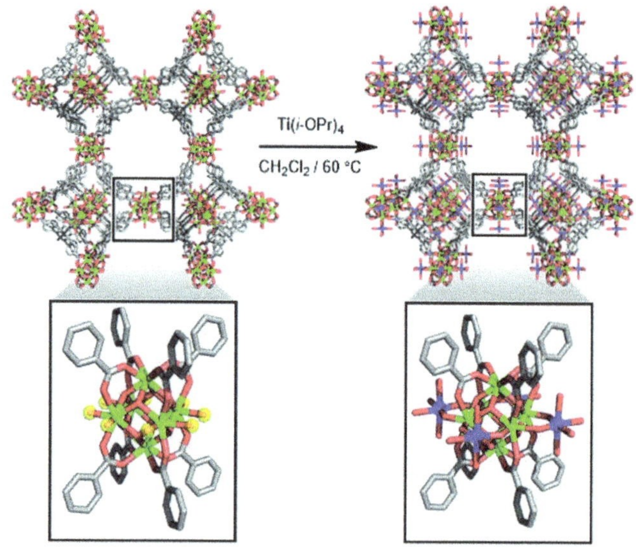

Figure 4. Synthetic scheme of Ti(IV) deposition on NU-1200, highlighting the exposed –OH/H_2O groups of the Zr6 cluster that are available to post-synthetic metalation. Green = Zr, Blue = Ti, Red = O, Gray = C. Adapted with permission from [54]; Published by WILEY, 2016.

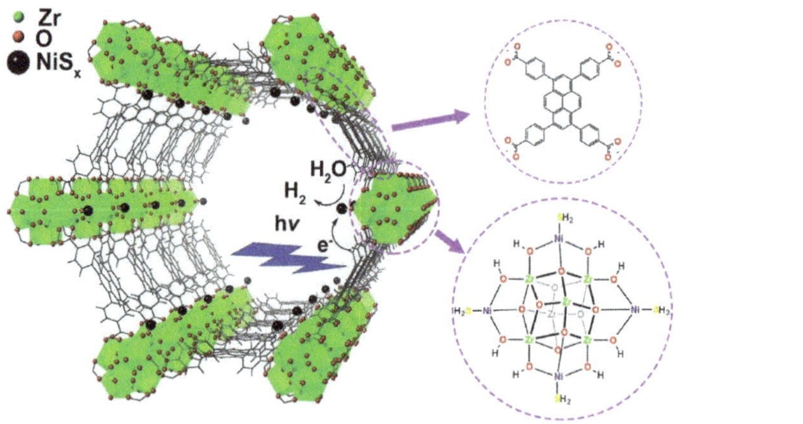

Figure 5. Idealized structure of NiS_x-functionalized NU-1000. The ligand and proposed structure of the atomic layer deposition in metal–organic frameworks (AIM)-metalated cluster are also shown. Green = Zr, Red = O, Black = NiS_x. Adapted with permission from [60]; Published by American Chemical Society, 2016.

An alternative to the utilization of the AIM strategy is through solution-phase metalation, sometimes referred to as solvent-assisted metal incorporation (SAMI) or solvothermal deposition in a MOF (SIM). In this technique, the preformed MOF is suspended in a solution of metal precursor for up to 24 h at relatively mild temperatures [54]. This technique is highly suited to stable Zr-MOFs, such as NU-1200 and NU-1000, as it is vital that the MOF be stable while suspended in solution at mildly elevated temperatures. As with AIM, metals are grafted onto the metal clusters via aqua/hydroxyl ligand pairs. A wide range of transition metals have been deposited on these two frameworks by Hupp, Farha, and Gagliardi, including Ti(IV), Mo(IV), Mo(VI), Nb(V), Co(II), and Cu(II) [62,65,66].

While a range of metal precursors can be used in SIM, a significant body of work uses with metal acetylacetonate species, due to their good solubility in organic solvents. In an early example of the versatility of this technique, the Farha, Hupp, and Nguyen groups functionalized UiO-66 with redox active V(V) ions. By treatment of UiO-66 with $VO(acac)_2$ (acac = acetylacetonate) in methanol, it was determined that V(V) ions could replace ~5% of the bdc linkers, binding through the -OH groups at the missing linker site [67]. While no crystal structures could be determined, the group proposed the binding mode for these compounds through a combination of NMR analysis of the digested MOF and the growth of additional peaks in an in situ DRIFTS spectrum. This functionalized MOF was then used to catalyze the oxidative dehydrogenation of cyclohexene [67]. Later work in the Gagliardi and Gates groups demonstrated the same principle utilized with vanadium for iridium complexes supported on the Zr_6 clusters of NU-1000 and UiO-66 [68]. Unlike the work from Farha, Hupp, and Nguyen, Gagliardi and Gates found that some of the ligands remained bound to the iridium center. No crystal structures were able to be determined for the iridium functionalization, however Gagliardi and Gates provided EXAFS (extended X-ray absorption fine structure) data with DFT (density-functional theory) calculations to assign the binding sites and modes of the iridium complexes. Later work demonstrated that these M-MOFs could be used as effective ethylene conversion catalysts [48].

Capitalizing on the principles outlined in the previous work in this field, several zirconium cluster systems as supports for earth-abundant transition metal catalysts have been investigated. One such example comes from the Lin group whose work primarily focused on post-synthetic metalation of the Zr_6 clusters in UiO-68. Deprotonation of the $Zr_3(\mu\text{-OH})$ sites with *n*-BuLi and treatment with $CoCl_2$ or $FeBr_2$ resulted in a M-MOF in which cluster H atoms were replaced by an organometallic catalyst moiety. The heterometals were bound to one oxide moiety from the cluster, both oxygen atoms of an attached linker, and a single halide atom in a distorted tetrahedral geometry [44]. Metalation of other zirconium cluster nodes with cobalt through μ_2 and μ_3 oxide binding was also studied. In all cases, the zirconium nodes acted as tridentate ligands to the heterometal [69,70]. These M-MOFs had high catalytic activity towards a wide range of organic reactions including: hydrogenation, borylation, amination, and silation [44,70,71]. This work has been expanded to consider alkali earth metal organometallic catalysts as well. In one such report of a UiO-69 structure, when the structure was treated with $MgMe_2$, highly decorated Zr_6 clusters could be achieved. While the MgMe bound to the clusters through deprotonated μ-OH moieties, the zirconium nodes here could also act as a monodentate ligand. This is because the coordination from the carboxylate functionality is very weak. After functionalization, the structure was demonstrated to maintain high activity for hydroboration and hydroamination catalytic reactions [70].

The Lin group later expanded their work to Ti-MOF cluster-based supports for cobalt catalysts. This work utilized the MIL-125 (MIL = Materials Institute Lavoisier) structure, which contains 12-connected $Ti_8O_8(OH)_4(COO)_{12}$ clusters. As with their previous work, it was demonstrated that deprotonation occurs at the μ-OH sites. Upon treatment of the structure with $CoCl_2$, insertion of a CoCl moiety into the center of the cluster was accomplished. Here the cobalt was bound through four μ-O moieties. An additional coordinated solvent molecule (THF, tetrahydrofuran) became bound to the cobalt center, creating a distorted octahedral geometry around the metal ion [72]. Upon treatment with $NaBEt_3H$, the reductive elimination of hydrogen resulted in Co-mediated electron transfer from the hydride, causing two of the eight Ti^{IV} centers to be reduced to Ti^{III}. This mixed valence M-MOF was then used to catalyze the hydrogenation of arenes and heteroarenes [72].

3.2. Post-Synthetic Metal Exchange

The composition of the clusters themselves can also be modified after the initial synthesis of the MOF through PSME. In general, this technique allows the metal clusters within the scaffold to be modified by substitution of existing metal ions with a different metal ion. This approach is advantageous as it allows the MOF to acquire additional functionality or stability through such substitution [73–75]. The major difference between PSME and post-synthetic metalation (PSM) is that in PSME, the metal ions in the clusters are exchanged with newly introduced metal ions yielding no net change in the total number of metal ions in a cluster. This is in contrast to PSM where metal ions are added to a cluster yielding a net change in the total number of metal ions in a cluster. In particular, the approach of PSME offers the ability to synthesize desired MOFs that are otherwise unable to be formed due to steric, electronic, or stability concerns during synthesis.

PSME refers to a range of techniques by which metal ions in a cluster are exchanged with a more active metal cation. Typically PSME is performed on clusters containing relatively inert metals, however, this technique is fairly versatile and can be performed on a wide range of metal clusters [76]. One of the earliest examples of PSME in the literature was in the replacement of cadmium(II) ions with lead(II) and lanthanide(III) ions in the anionic MOF $[(Cd_4O)_3(hett)_8]^{6-}$ (H_3hett = ethyl substituted truxene tricarboxylic acid). Complete substitution was achieved by soaking the preformed MOF in aqueous lead nitrate or a lanthanide salt solution for two days. The partial substitution for this exchange could be achieved using shorter reaction times. The single-crystal to single-crystal transformation for this system was demonstrated to be completely reversible under mild conditions [77].

An important advantage for PSME is that it can be used to generate MOFs which are unable to be synthesized in a particular topology. Recent work in the Volkmer and Dinca groups has shown the usability of PSME for zinc MOF metal substitutions. MFU-4*l* (MFU = metal–organic framework, Ulm University), a zinc MOF with five zinc(II) cations in the cluster, is of topological and catalytic interest, but cannot be synthesized in situ with other metals beyond zinc [73,78]. Despite being unable to directly synthesize the isostructural MOFs of MFU-4*l* with other metals, PSME can be accomplished for the peripheral zinc ions with titanium, chromium, and cobalt (Figure 6) [73,75,78]. Substitution of the zinc ions with more active heterometal ions increased the activity of the MOF for a variety of catalytic applications, including: CO oxidation, olefin polymerization, and olefin insertion [73].

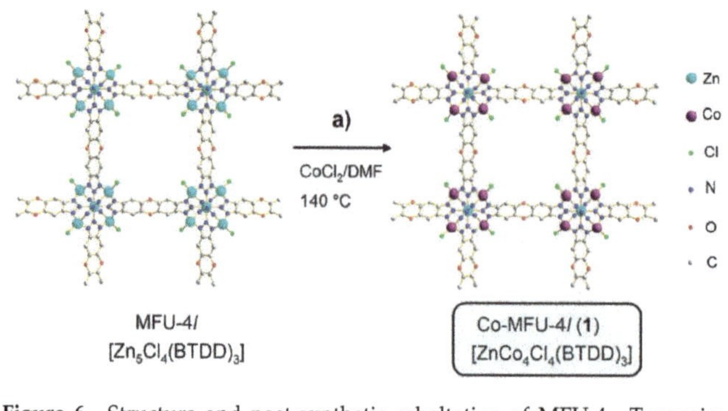

Figure 6. Structure and post-synthetic cobaltation of MFU-4. Turquoise = Zn, Maroon = Co, Green = Cl, Blue = N, Red = O, Gray = C. Adapted with permission from [78]; Published by Royal Society of Chemistry, 2012.

Similar to the work demonstrated in MFU-4*l*, substitution of inert nickel ions with active cobalt ions in Ni-MOF-74 has also been demonstrated. The resultant M-MOF demonstrated a higher catalytic conversion of cyclohexene than either pure Ni-MOF-74 or Co-MOF-74. This indicates the potential for high catalytic activity in M-MOFs in comparison to their homometallic MOF counterparts, per mole of active metal. It is believed that the increased catalytic activity demonstrated in this example stems from substitution of only the most accessible metal ions in the MOF [79].

PSME can also be utilized when the direct synthesis of a desired MOF cannot be accomplished in a predictable way. Titanium MOFs are highly desired as the titanium clusters within the MOF mimic the chemistry of titanium nanoparticles. Titanium is also a low toxicity earth abundant metal. Although titanium MOFs are highly desired for catalytic applications, it is often not possible to directly synthesize Ti-MOFs in a predictable way as the modulation techniques used for the analogous Zr- and Hf-MOFs do not work for titanium structures [45]. One highly accepted approach to the synthesis of Ti-MOFs is through the PSME of preformed Zr- or Hf-MOFs.

Cohen's group reported a total metal replacement of up to 38% of the Zr(IV) cations with Ti(IV) cations could be accomplished for UiO-66. The resulting structures retained the crystallinity of the original material, and have been shown to be both highly porous and stable [76]. However, a later study by the same group disproved this analysis, as scanning tunneling electron microscopy revealed that the Zr(IV) cations were not in fact substituted. Instead, a nanoscale TiO_2 layer was deposited on the surface of the particles, suggesting the exterior surface of the MOF was post synthetically metalated and that no cation exchange occurred. The Zr(IV) clusters in the interior of the crystal remained unaffected by the surface metalation and no zirconium was observed in the supernatant by ICP-MS (inductively coupled plasma mass spectrometry) following the supposed exchange [49].

Despite this setback, the Zhou group demonstrated that post-synthetic titanation is in fact possible and is not limited to substitution within preformed group IV MOFs. Their work demonstrated that replacement of metal ions such as scandium(III), zinc(II), and magnesium(II) with titanium(IV) in MOFs through the use of a Ti(III) intermediate can also be accomplished. By submersion of the preformed MOF in a Ti(III) rich solution, Ti(III) can be incorporated into the MOF structure and oxidized to yield Ti(IV). This can be accomplished without destroying the crystallinity or altering the overall structure of the parent MOF. However, judicious selection of the host MOF is crucial. The host MOF must be stable towards the Ti(III) intermediates, contain open metal sites and labile linkers to accelerate the metathesis, and contain a coordination site that will readily accommodate the substituted metal [80]. A similar technique has been used to replace magnesium with Fe(II) and Cr(II).

Other highly redox active metals such as iron(II/III) have also been investigated for PSME to improve the catalytic activity of the manganese(II) metal clusters. Using iron(III) chloride as an iron source, near complete substitution of manganese ions in Mn_3L_3 (H_2L = bis(4-(4′-carboxyphenyl)-3,5-dimethyl-pyrazolyl)methane) was performed [81]. M-MOFs containing smaller fractions of iron were also obtained by using shorter reaction times, lower temperatures, and alternative iron sources. The similarity between the parent Mn_3L_3 structure and the resulting M-MOF suggests that the iron incorporated in the structure is iron(II), despite coming from an iron(III) source. Most likely, the change in oxidation state is due to the original Mn clusters acting as a reductant [81].

3.3. *In Situ Cation Doping*

In situ cation doping is a less common, one-pot method for cationic substitution in MOFs as compared to post-synthetic methods. Utilizing this method, it is significantly more difficult to perform and direct the desired structure change in the M-MOF [82]. Cation doping generally involves partial substitution of the metal source with a heterometallic source. This strategy leads to a M-MOF in which the cations in the clusters are partially substituted with the heterometallic cation. This approach can be utilized for a one-pot alteration of a parent MOF in regard to enhanced activity [83] or stability [82,84].

One well known example of in situ cation doping is exemplified in the Long group's work on iron-oxo species within MOFs. The Long group used a Fe-MOF for the conversion of ethane to ethanol using NO_2 under mild conditions. Although the iron sites are critical for the desired chemical conversion, it was necessary to substitute a large fraction (~95%) of the iron sites with redox-inactive magnesium ions. This was done to prevent oligomerization and overoxidation of the substrate. Rather than performing this substitution post-synthetically, the M-MOF was created by partial substitution of the iron source with a similar magnesium source during the solvothermal synthesis [83]. The oxidation state uniformity for the iron sites exhibited in the Mössbauer spectrum of the material indicated that the magnesium substitutions were distributed evenly throughout the structure. Although the final M-MOF was successful in catalyzing the oxidation of ethane to ethanol, it also suffered significant catalyst deactivation issues [83].

In situ cation doping has also been shown to be an effective technique for the increase in capture affinity of a desired substrate or to increase the stability of a desired MOF. In this work by the Shi and Cheng groups, MOF-5 was doped with either cobalt or nickel, demonstrating the improved H_2 uptake capabilities of each respective MOF as compared to the parent MOF-5 [84,85]. In addition to improving the uptake performance, it was also noted that for the heterometallic MOFs, the hydrostability of MOF-5 could be improved. A key drawback to MOF-5 is its instability, as the Zn–O bonds of MOF-5 and other related benzyl chloride MOFs can be hydrolyzed by atmospheric humidity [84,86–88]. While MOF-5 begins to degrade within two days of exposure to atmospheric humidity, Ni22-MOF-5 (22% Ni-doped) remained stable for more than one week (Figure 7) [84]. Ni-doped MOF-5 could also be combined with reduced graphene oxide in order to produce a high capacity energy storage composite material [82].

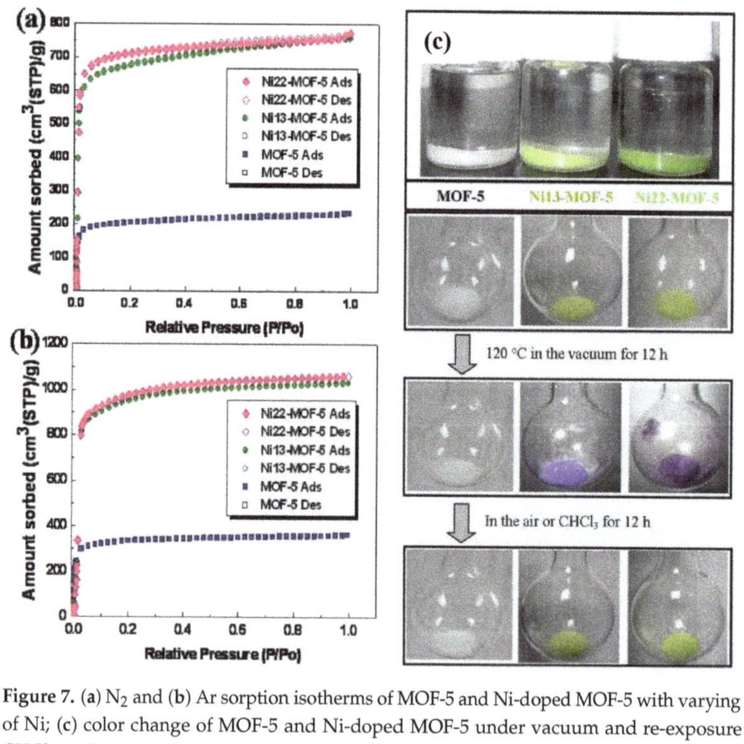

Figure 7. (**a**) N_2 and (**b**) Ar sorption isotherms of MOF-5 and Ni-doped MOF-5 with varying amounts of Ni; (**c**) color change of MOF-5 and Ni-doped MOF-5 under vacuum and re-exposure to air of $CHCl_3$. Adapted with permission from [84]; Published by American Chemical Society, 2012.

4. Organometallic Catalysis in MOFs

One of the most important advances for the industrial scale production of chemicals is the incorporation of catalysts into chemical processes. By incorporating organometallic compounds, great developments in our modern world have taken place, particularly for oxidation, hydrogenation, C–H activation, and coupling reactions.

Although organometallic compounds have revolutionized the way we make and tailor chemical products, there are still some major drawbacks to organometallic compounds as catalysts for chemical processes. Many organometallic catalysts have limited lifetimes, which is most often caused by deactivation of the active site. Deactivation can occur through several pathways such as M–L redistribution, M–L oligomerization, and M–L conversion [89]. In addition, when used on an industrial scale, many organometallic catalysis are not recoverable from the reaction media as they are difficult to separate from the products in post-production. This decreases the overall economy of production.

The structure and tunability of MOFs provide a potential way to immobilize organometallic catalysts within the framework. This is advantageous as it increases both catalytic performance and the turnover number by protecting the catalyst. The added feature of the incorporation of heterogeneous catalysts within the framework also helps to prevent some common problems in homogeneous systems [90]. Here, we introduce several pieces of work that compare the performances of organometallic catalysts incorporated into MOFs to their free variants.

4.1. Reactions of Alkenes

Alkenes are very important chemicals for energy and chemical synthesis. Many important industrial alkene reactions, such as hydrogenation, oxidation, and isomerization reactions incorporate noble metals as catalysts [91,92]. For example, a catalyst that has been broadly incorporated into hydrogenation reactions for olefins, Wilkinson's catalyst, is Rh-based. Other noble metals, such as Pd have also been incorporated for olefin hydrogenation and oxidation reactions. These metal centers have been utilized on a large scale for alkene reactions but are very expensive and there is a limited quantity available. MOFs have been demonstrated to be highly efficient, size and conformationally selective catalysts even without the incorporation of noble metals as catalytic centers. The potential of utilizing MOFs as catalysts for reaction of alkenes is to take advantage of the properties of the material such as high specific surface areas, functionalized selectivity, and high gas sorption while also eliminating the need for noble metals in the catalytic conversion. Many of the reported works also realize highly efficient conversions of alkenes.

Site-isolation of the active sites within the MOF scaffold is crucial to keep active centers active for a prolonged lifetime. A recent example of this strategy was reported by the Lin group. In their work, TPHN-MOF-MgMe (TPHN = 4,4′-bis-(carboxyphenyl)-2-nitro-1,1′-biphenyl) was prepared by introducing MgMe sites onto the μ_3-OH groups in TPHN-MOF's nodes (Figure 8). As a result of the well-separated Mg-species, the Schlenk-type ligand redistribution was successfully avoided. This MOF was then later successfully utilized for catalytic hydroboration of imines and hydroamination of amino alkenes under mild condition [93].

In addition to maintaining site activity, MOF scaffolds can be rationally designed to adjust the site specificity of the active-sites and inner structure. In this way, the framework can show improved performance as a kind of micro-reactor for chemical conversions. For instance, the Zhou group reported a one-step synthesis of a core-shell MOF, PCN-222(Fe)@Zr-BPDC (BPDC = biphenyl-4,4′-dicarboxylate), that enables the catalytic epoxidation of alkenes [94]. This hybrid core-shell structure was guided by differences in the nucleation rates of PCN-222′s fast homogeneous nucleation and Zr-BPDC's accelerated heterogeneous nucleation. Its remarkable size-selectivity is studied through substrate expansion, which shows that the narrow open window will increase the diffusion resistance, thus block the reaction pathway for large-size alkenes like (Z/E)-1,2-diphenylethene. Moreover, Eddaoudi and co-workers successfully anchored $W(\equiv C^tBu)(CH_2{}^tBu)_3$ onto

Zr_6 clusters of NU-1000 [95]. The resulted material featured high catalytic properties in olefin metathesis.

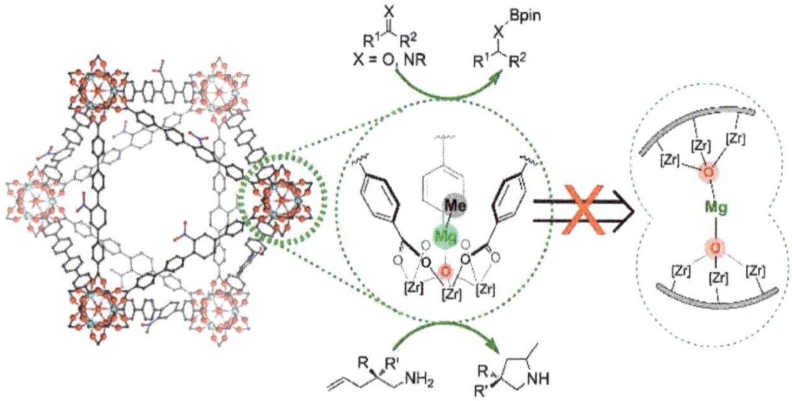

Figure 8. The site-isolation strategy of introducing MgMe sites as alkenes activation center. Red = O, Black = C, Blue = N, Green = Zr. Adapted with permission from [93]; Published by American Chemical Society, 2016.

In another study, trans-coordinate metal active sites were introduced into a MOF scaffold through sequential ligand elimination and installation. This trans-coordinate M-INA_2 (INA = 4-pyridinecarboxylate) MOF demonstrated higher activity in ethylene dimerization reactions over any known cis-binding MOF variant. [40] The trans-binding site approach, although higher in activity, is much more difficult to generate than a cis-binding site variant. A trans-coordinated site was synthetically accomplished using PCN-160 as a start material and simultaneously replacing the parent azobenzene-4,4′-dicarboxylate (AZDC) linkers with 4-carboxybenzylidene-4-aminobenzate (CBAB) linkers [96]. The CBAB linkers were then exchanged with M-INA_2 linkers. Through this multi-step ligand exchange, a MOF with trans-coordinated sites in the linkers was successfully synthesized. Neither a direct synthesis of the M-INA_2 containing PCN-160 or the exchange of the original AZDC ligands with M-INA_2 was unable to yield the desired MOF. The trans-chelated sites within this MOF demonstrated a significantly higher catalytic efficiency over the *cis*-binding sites previously demonstrated in any other MOF.

4.2. *Pd-Catalyzed Cross-Coupling Reactions*

Palladium-catalyzed cross-coupling reactions are considered to be one of the greatest breakthroughs in the field of organic chemistry in the modern era. This type of reaction is so important that in 2010 the Nobel Prize in Chemistry was awarded jointly to Drs. Richard F. Heck, Ei-ichi Negishi, and Akira Suzuki "for palladium-catalyzed cross coupling in organic synthesis." This method is highly efficient for building carbon-carbon and carbon-nitrogen bonds directly, in high selectivity. These reactions also operate at relatively low temperatures and are often very high yielding [97].

After a long period of development, the usage of palladium-catalyzed coupling reactions to produce fine chemicals has significantly increased, but there are still some downsides of using the traditional catalytic protocols for industrial scale synthesis. For example, when performing the Heck coupling reaction, using ligands can prevent aggregation but it will make separation of the final products more difficult. If using palladium nanoparticles, the reaction will result in the catalyst precipitating after the reaction [98]. Immobilizing Pd species in MOF scaffolds can be utilized to transplant these Pd-catalyzed reactions into heterogeneous system [57].

NHC ligands can be utilized to fixate Pd catalysts within MOF scaffolds. One such example of this, reported by the Dong group, was in the case of Pd(II)-NHDC-UiO-67

(Figure 9). This MOF was developed for the catalytic Heck cross-coupling reaction of benzyne-benzyne-alkenes [99]. This approach demonstrated not only good catalytic efficiency for a broad range of substances, but also was highly recyclable. Another case of the NHC-MOF approach is in NH_2-MIL-101(Cr). In this work, the framework was utilized to catalyze Suzuki-Miyaura cross-coupling reactions for a variety of substrates [100]. There have also been reports of Pd incorporated MOFs that utilize other ligand beyond the NHC ligand approach. For example, the Li group reported a bpdc (para-biphenyldicarboxylate) ligand-based UiO-67 framework to immobilize Pd [101] and the Cohen group used a bpy ligand to chelate a Pd sites on the same parent skeleton [102]. Both approaches were shown to be versatile in that each could catalyze multiple Pd-catalyzed cross-coupling reactions.

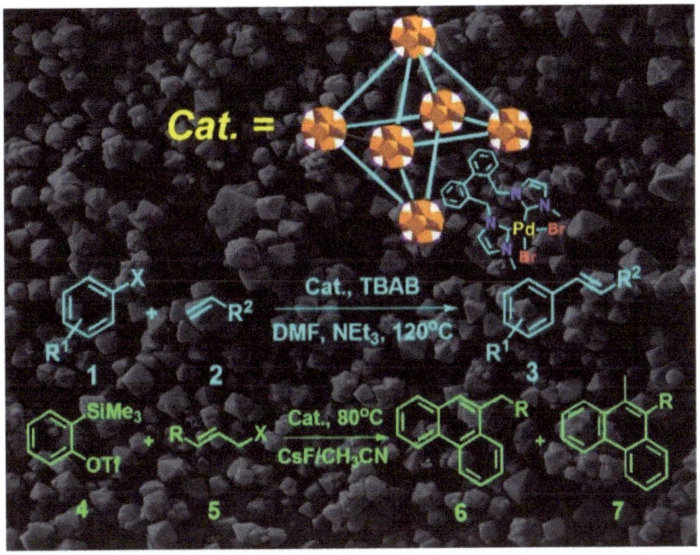

Figure 9. NHC (*N*-heterocyclic carbene) ligand bonded Pd sites in MOFs for Heck cross-coupling reaction. Orange = Zr-O clusters. Adapted with permission from [99]; Published by American Chemical Society, 2018.

An interesting application for Pd incorporated MOFs is in their use as sensors for Cu(II) ions, this application was first reported by the Jiang group [103]. In their work, they doped Pd sites into PCN-222 to form PCN-222-Pd(II) as a sensor. When copper ions were present in solution, the strong binding affinity of the porphyrin centers in the MOF readily replaced the Pd centers with Cu. The Pd ions were leached into the framework and were then reduced into nanoparticles. The PCN-222 framework was able to stabilize the nanoparticles which could be utilized as catalysts. The freshly generated Pd NPs were demonstrated to be efficient in catalytic *N*-Allyl-2-iodoaniline Heck cross-coupling reactions. The substrate was easily converted into 3-methylindole by the catalyst. As 3-methylindole has strong fluorescence, this system could be quantitative for the concentration of Cu(II) ions in solution [103].

4.3. C–H Activation Reactions

C–H activation reactions are chemical conversions that transform relatively inert C–H bonds into C–X bonds (X can be C, O, N and so on) [104]. These reactions have great value to petrochemical conversions as they can simplify the process of synthesizing desired chemicals from cruder starting materials [105]. The greatest challenge for these reactions involves the relatively high stability of saturated C–H bonds and the issue of selectivity [106].

Traditionally, there are three main approaches for C–H activation. The first approach to initiation is the removal of a hydrogen atom by using either a base, radical, or electrophile and subsequent attack of the active carbon atom. In the second approach an oxidative reaction is used to activate the carbon atom. This is the most widely used approach. In this approach, a metal atom inserts itself into the C–H bond to activate the carbon atom. The third approach for C–H activation is derived from metal-oxo enzyme chemistry. In this approach, a metal-oxo complex reacts with the hydrogen atom in the C–H bond forming a hydroxyl group. As a result, the carbon becomes activated [107]. Many of the reported work on C–H activation use organometallic compounds and are often second and third row transition metal-based. Metals such as Ru, Rh, Ir, and Pd are good candidates for the C–H active metal centers. The vast majority of these reactions are still performed in the homogeneous catalytic fashion [108]. As a result, it is feasible to assist in eliminating several common drawbacks while maintaining high performance of the active sites through using MOFs as the platform for this catalysis [109].

One example of the utilization of MOFs for C–H activation comes from the Lin group who reported the use of UiO-Co and UiO-Fe for C–H borylation, silylation, and amination in 2016 [70]. In their work it was observed that the reaction proceeded through deprotonation of $Zr_3(\mu_3\text{-}OH)$ in the nodes of UiO-68 by nBuLi. The activated catalyst then reacted with a $CoCl_2$ or $FeBr_2{\cdot}2THF$ precursor, forming the active species μ_4-O-M-H (M = Co or Fe) [44]. In the same year, the Lin group also reported a linker-based (mono)phosphine-M complex active site, P_1-MOF·M (M = Rh and Ir), which was shown to catalyze the C–H borylation of arenes [110]. The active M-PR_3 complexes were able to be stabilized through the isolation of catalytic sites in the frameworks. Another approach published by the Zhou group (Figure 10) shows the possibility of postsynthetic incorporation of active sites into MOF's pores to realize C–H activation [111]. In this work an NNN-pincer-based PCN-308 MOF scaffold, synthesized from 4′-(4-carboxyphenyl)-[2,2′:6′,2″-terpyridine]-5,5″-dicarboxylic acid(H_3TPY) and Zr_6 clusters (Figure 10), was shown to be effective for C–H borylation reactions. This MOF had metal ions postsynthetically introduced to the pincer ligands forming highly active M@PCN-308, where M=Co being exceptionally effective.

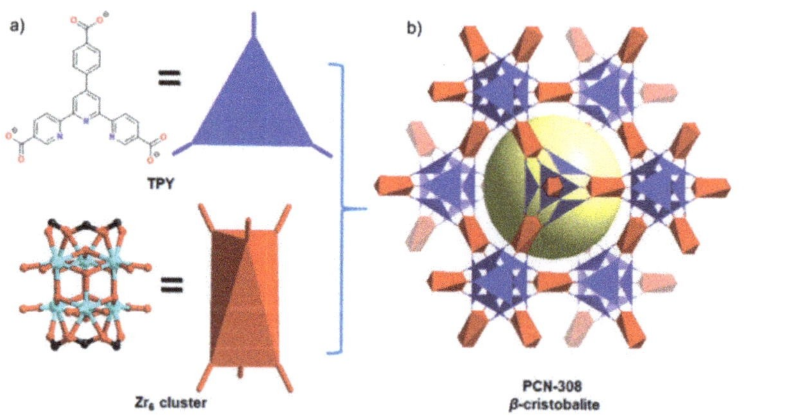

Figure 10. (**a**) Trigonal-planar organic linker TPY and six-connected D_{3d}-symmetric Zr_6 antiprismatic clusters. (**b**) The β-Cristobalite network of PCN-308 simulated based on the reported PCN-777 structure. Blue = TPY ligand, Red = Zr_6 cluster, Yellow = void space. Adapted with permission from [111]; Published by Royal Society of Chemistry, 2019.

5. New Horizons to the Use of MOFs for Organometallic Transformations

Owning to their high tunability and versatile pore environments, MOFs are promising candidates for applications that require the incorporation of diverse function groups within highly ordered molecular systems [112]. The combination of utilizing organometallic chemistry within the MOF frameworks can also introduce and improve the properties of MOFs for designer applications. In this section we will explore how the incorporation of organometallic chemistry into MOF frameworks can be utilized for real world applications beyond traditional organometallic applications. This work will only focus on gas sorption and separations, spin ordering in magnetism, and therapeutics, although there is a vast array of potential applications for these materials.

5.1. Gas Adsorption

In organometallics, metal centers with unsaturated coordination sites can possess high activities and strong interactions with specific gas molecules. For example, heme, the metal-porphyrin found in human blood and muscle tissue, contains a coordinately unsaturated Fe center. This Fe center is responsible for the controlled capture and release of O_2 in hemoglobin and myoglobin. The controlled binding and release of specific gases can also be applied in application-based design of MOFs. When the metal centers within the MOF framework are designed with open metal sites and their electronic environment is tailored to a specific selectivity, the metal sites can selectively and reversely bind specific gas molecules. For instance, in 2014, the Harris group observed binding between O_2 and a four-coordinate ferrous center in a porphyrin-based MOF, PCN-224(Fe). Their study generated a superoxide Fe^{III}- porphyrin at −78 °C [113]. In later works, this system was be further extended to PCN-224(Mn) which also demonstrated selective reversible O_2 binding [114].

Moreover, through tuning the ligand environment and metal site electronics, MOFs with open metal sites can be synthesized to achieve selectivity for a desired gas. For instance, the Dincă group once observed NO disproportionation on an $[FeZn_3O]$ center in a partial cation exchanged Fe^{2+}-MOF-5 [115]. In 2015, the Long group reported a MOF, Fe_2(dobdc) (dobdc = 2,5-dioxido-1,4-benzenedicarboxylate), which contained coordinately unsaturated redox-active Fe^{2+} centers. Upon exposure to NO gas, a $Fe_2(NO)_2$(dobdc) species was formed [116]. This reversable binding involved the formation of Fe^{3+}-NO^- adducts, which could be tuned for the gradual release of NO gas. In another example from 2016, the Dincă group designed a series of MOFs, $M_2Cl_2(BTDD)(H_2O)_2$ (M = Mn, Co, Ni, BTDD = bis(1H-1,2,3-triazolo[4,5-b],[4′,5′-i])dibenzo-[1,4]dioxin), that exhibited abundant, well-dispersed open metal sites with a highly selective uptake of NH_3 [117]. Building upon their previous work and that of the Dincă group, in 2017 the Long group reported two similar Fe-MOFs, Fe_2Cl_2(BTDD) and Fe_2Cl_2(BBTA) (BBTA = 1H,5H-benzo(1,2-d:4,5-d′)bistriazole), containing unsaturated Fe^{II} centers (Figure 11) [118]. Upon exposure to CO, the neighboring Fe centers along the coordination chain underwent a high-spin to low-spin transition, leading to cooperative adsorption of CO with very low regeneration energies. The selective CO scavenging through the Fe binding site spin-state transition could also be observed in the MOF, $Fe_3[(Fe_4Cl)_3(BTTri)_8]_2 \cdot 18CH_3OH$ (BTTri = 1,3,5-tris(1H-1,2,3-triazol-5-yl)benzene) [119]. The Long group synthesized a MOF Fe_2(dobdc) (dobdc: 2,5-dioxido-1,4-benzenedicarboxylate) with open Fe(II) sites, which can coordinate with olefins reversibly [120]. These properties were applied in membrane-based separation of olefin/paraffin.

Remarkably, redox-active properties at the metal site can also be applied for the capture of gas molecules. Some low-valence metals can form reversible binding with oxidant gas to give a high valence, while the robustness of the frameworks can still be maintained. In 2017, the Dincă group generated a MOF with reversable binding between Co_2Cl_2(BTDD) and the elemental halogens Cl_2 and Br_2 [121]. In this system, the Co^{II} centers in the Co_2Cl_2(BTDD) MOF became oxidized by Cl_2 or Br_2 yielding a Co^{III} center. This process was quantitative and demonstrated the reversibility of the bond through the release of the halogen molecules while heating. A radical mechanism is proposed to explain the homolytic cleavage of the $Co^{III}-X$ bond to give halogen elements.

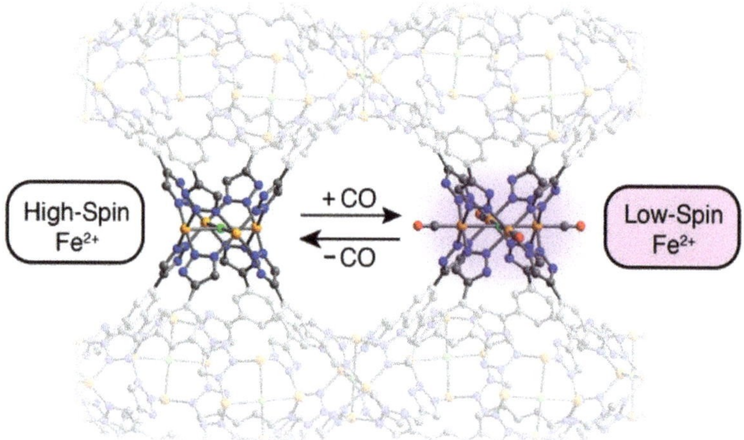

Figure 11. Reversible CO scavenging observed in MOF $Fe_3[(Fe_4Cl)_3(BTTri)_8]_2 \cdot 18CH_3OH$ (BTTri = 1,3,5-tris(1H-1,2,3-triazol-5-yl)benzene) driven by spin-state transition in Fe centers. Blue = N Red = O. Yellow = Fe. Green = Cl. Grey = C. Adapted with permission from [119]; Published by American Chemical Society, 2016.

5.2. Magnetism

Magnetism is one of the primary properties of organometallic compounds and porous magnets have drawn great attention owning to their potential applications in high-density information storage. The low working temperatures of many magnets, including porous magnets, restrict their practical application. For instance, the record high ordering temperature of magnetic MOFs was only 32 K in 2014 [122]. It is usually contradictory to design a MOF with a high ordering temperatures and large surface area. This is because the super-exchange coupling between the paramagnetic centers decreases dramatically with increased distance between centers. While ligands typically used in MOF synthesis seem to be too long to give strong coupling between magnetic centers, some groups in recent years have reported cyan-functionalized ligands as possible solutions to this problem. In particular, TCNE and TCNQ (TCNE = tetracyanoethylene, TCNQ = 7,7,8,8-tetracyanoquinodimethane) have been used to construct network magnets with voids and high ordering temperatures. However, the weak coordination bonds between these ligands and the transition metal centers make it difficult to obtain robust permanent pores in the structures [123–125]. Such an issue had not been solved until 2015, when the Harris group utilized a semiquinoid ligand, 2,5-dichloro-3,6-dihydroxy-1,4-benzoquinone to synthesize a radical-bridging MOF, $(Me_2NH_2)_2[Fe_2L_3] \cdot 2H_2O \cdot 6DMF$ (DMF = dimethylformamide), that had an ordering temperature of 80 K and a BET surface area of 885 m^2/g [126]. During the synthesis of this MOF, two thirds of the ligands could be reduced into a radical form. This unique property allowed for bridging between adjacent Fe^{III} centers through strong antiferromagnetic coupling. Further studies indicated that once the ligands were fully reduced to their radical form, the ordering temperature of the MOF could be increased to 105 K [127].

5.3. Quantum Computation

Qubits, objects existing in any quantum superpositions of two states, are the fundamental units of quantum information systems. An individual paramagnetic compound can be utilized as a qubit due to the diversity in spin states that can exist in many transition metal complexes. The spin can be tuned from a spin up to a spin down, or visa-versa, allowing for precise control of the qubit spin state. A spatial organization of qubits into an array is a promising strategy for the construction of quantum sensors and information processing systems. Unfortunately, qubits have a downside. They are very sensitive to magnetic noise, ultimately impeding current development in this field. To overcome the barrier, in 2017, the Freedman group fabricated qubits into the framework of PCN-224 by mixing paramagnetic Co^{II}-porphyrin and diamagnetic Zn^{II}-porphyrin to achieve $[(TCPP)Co_{0.07}Zn_{0.93}]_3[Zr_6O_4(OH)_4(H_2O)_6]_2$ [128]. The qubits in the framework exhibited a clock transition, thereby eliminating the influence of local magnetic noise, yielding a longer lifetime of 14 μs for the qubits (Figure 12). Building upon this work, in 2019, the same group constructed a PCN-224 framework that was fully substituted with Cu^{II}-porphyrin to achieve concise position control of the qubits [129]. In this system, a precise array of qubits, which were Cu^{II} sites on porphyrinic ligands, could be achieved by controlling the distances between Cu^{II} sites. This work not only showed that vibrational environments in the lattice could affect the spin dynamics of the qubits, but it also provided a pathway to construct qubit arrays precisely.

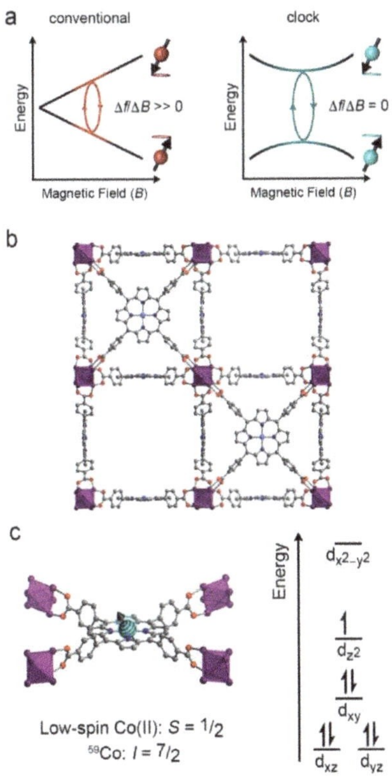

Figure 12. (**a**) Depiction of the clock concept for electronic spin qubits. Clock EPR transitions possess frequencies (*f*) that are insensitive to small changes in magnetic field (ΔB). This property may enable a long T2 in magnetically noisy environments relative to a conventional EPR transition. (**b**) Crystal structure of the parent $[(TCPP)\text{-}Co]_3[Zr_6O_4(OH)_4(H_2O)_6]_2$ framework (TCPP = tetrakis(4-carboxyphenyl)-porphyrin) [128]. This material is porous, exhibiting a high surface area (~3200 m^2/g), and hosts paramagnetic cobalt(II) metal ions. (**c**) Close-up view of the square-planar cobalt(II) ions in the structure and qualitative depiction of the d-orbital splitting. The cobalt(II) ions are low-spin, featuring one unpaired electron that is strongly coupled to the I = $^7/_2$ ^{59}Co nucleus. This strong hyperfine interaction engenders an avoided crossing such as that shown in (**a**). Purple = Zr. Cyan = Co. Grey = C. Red = O. Adapted with permission from [128]; Published by American Chemical Society, 2017.

5.4. Therapeutic

Numerous organometallic compounds have been shown to have a variety of applications in medicine. For example, cisplatin is an organometallic chemotherapy drug derived from platinum. Although many organometallic compounds are highly effective for medicinal applications, they are often difficult to direct to precise locations within the body, leading to unwanted side effects. Owning to intrinsic porosity and surface tunability, nanoscale MOFs can also serve as drug delivery agents to improve location targeting for an organometallic complex within the body. Some of the seminal work in this field came from the Lin group who has been working on a systematic study on this field. In 2009, they loaded a cisplatin prodrug, ethoxysuccinato-cisplatin (ESCP), onto NH_2-BDC ligands of nanosized MIL-101(Fe) (NH_2-BDC = 2-aminoterephthalic acid) [130]. The bonding between ESCP and NH_2-BDC was quite labile causing a more gradual release of the ESCP. In 2014, the Lin group reported a successful loading of the cisplatin prodrug onto a Zr-MOF. The MOF@drug complex could be further coated with small interfering RNA, improving the chemotherapeutic efficacy [131].

Specific organometallic compounds can generate active agents to inhibit cancer cell growth. In 2018, Lin group constructed a cationic nanoscale MOF, Hf-DBB-Ru (DBB-Ru = bis(2,2′-bipyridine)(5,5′-di(4-benzoato)-2,2′-bipyridine)-ruthenium(II) chloride), that featured a high mitochondrial targeting ability. In their study, it was found that DBB-Ru could generate singlet oxygen and the Hf clusters could produce hydroxyl radicals once exposed to X-ray during radiodynamic therapy and radiotherapy [132]. Such a strategy was further extended to Hf-MOFs loaded with photosensitive $Ir(bpy)[dF(CF_3)ppy]_2^+$ ligands. This combination within the scaffold yielded instant and reproductive death of tumor cells even at a modest X-ray dose (Figure 13) [133]. In another example, the Furukawa group installed metal carbonyl ligands onto UiO-67, achieving CORF-1, $MnBr(bpydc)(CO)_3$@UiO-67 (bpydc = 5,5′-dicarboxylate-2,2′-bipyridine). CORF-1 was demonstrated to be able to release CO under exposure to light and have great potential in therapy for several inflammatory diseases [134].

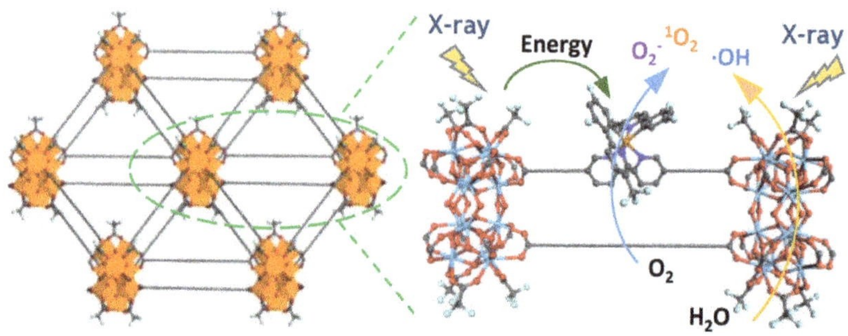

Figure 13. A combination of Hf_{12} cluster and photosensitive ligand featuring enhanced behavior in radiotherapy and radiodynamic therapy. Orange = Hf-oxo clusters. Red = O. Cyan = Hf. Blue = N. Green = F. Grey = C. Adapted with permission from [133]; Published by American Chemical Society, 2018.

6. Conclusions

The purpose of this review is to showcase the ultra-high tunability, versatility, and originality of MOFs as a next-generation approach to organometallic catalytic processes. Due to the structural features and diverse utility of these structures, MOFs are emerging as premier catalytic systems for the chemical transformations of materials traditionally only converted using organometallic catalysts. From the linkers within the frameworks, we are able to design specific catalytic reactions for site-isolated catalysts with constrained systems not possible in traditional directions. These features offer a unique study of the impacts of ligand steric hindrance effects as well as the influences of electronics of catalytic metal centers. The metal center functionalization offers a unique approach to the tunability of a system towards desired chemical outcomes. As a result, both ligand-based and metal cluster-based studies have aided in the development of MOF centered organometallic chemical processes. As new and diverse structures of MOFs are developed, a greater understanding of the potential applications of these systems becomes available. MOFs offer a promising alternative for an even brighter future through their utilization in organometallic chemical processes. We expect that given the relative infancy of this field to that of catalytic chemical transformations, we will see many new approaches to this important area of research in the foreseeable future.

Author Contributions: Conceptualization, H.F.D., L.F., H.-C.Z.; writing—original draft preparation, F.C., J.A.P., K.-Y.W., T.-H.Y., H.F.D.; writing—reviewing and editing, F.C., H.F.D., L.F.; visualization, K.-Y.W.; supervision, H.-C.Z.; project administration, H.F.D., F.C.; funding acquisition, H.-C.Z. All authors have read and agreed to the published version of the manuscript.

Funding: This work was supported by the Robert A. Welch Foundation through a Welch Endowed Chair to H.-C.Z. (A-0030).

Data Availability Statement: Data sharing not applicable.

Acknowledgments: The authors acknowledge the financial supports of Robert A. Welch Foundation through a Welch Endowed Chair to H.-C.Z. (A-0030).

Conflicts of Interest: The authors declare no conflict of interest.

References

1. Whitesides, G.M. Reinventing chemistry. *Angew. Chem. Int. Ed.* **2015**, *54*, 3196–3209. [CrossRef] [PubMed]
2. Barnaby, R.; Liefeld, A.; Jackson, B.P.; Hampton, T.H.; Stanton, B.A. Effectiveness of table top water pitcher filters to remove arsenic from drinking water. *Environ. Res.* **2017**, *158*, 610–615. [CrossRef] [PubMed]
3. Benavides, P.T.; Sun, P.; Han, J.; Dunn, J.B.; Wang, M. Life-cycle analysis of fuels from post-use non-recycled plastics. *Fuel* **2017**, *203*, 11–22. [CrossRef]
4. Phalke, V.S.; Arote, A.P.; Sonawane, R.R. Design and development of machine which generates fuel using pyrolysis of waste plastic. *Int. J. Sci. Res. Dev.* **2017**, *5*, 1–6.
5. Li, H.; Eddaoudi, M.; Groy, T.L.; Yaghi, O.M. Establishing microporosity in open metal-organic frameworks: Gas sorption isotherms for Zn(BDC) (BDC = 1,4-benzenedicarboxylate). *J. Am. Chem. Soc.* **1998**, *120*, 8571–8572. [CrossRef]
6. Yaghi, O.M.; Li, G.; Li, H. Selective binding and removal of guests in a microporous metal-organic framework. *Nature* **1995**, *378*, 703–706. [CrossRef]
7. Kirchon, A.; Feng, L.; Drake, H.F.; Joseph, E.A.; Zhou, H.-C. From fundamentals to applications: A toolbox for robust and multifunctional MOF materials. *Chem. Soc. Rev.* **2018**, *47*, 8611–8638. [CrossRef]
8. Dhakshinamoorthy, A.; Garcia, H. Metal–organic frameworks as solid catalysts for the synthesis of nitrogen-containing heterocycles. *Chem. Soc. Rev.* **2014**, *43*, 5750–5765. [CrossRef] [PubMed]
9. Dhakshinamoorthy, A.; Santiago-Portillo, A.; Asiri, A.M.; Garcia, H. Engineering UiO-66 metal organic framework for heterogeneous catalysis. *ChemCatChem* **2019**, *11*, 899–923. [CrossRef]
10. Dhakshinamoorthy, A.; Asiri, A.M.; García, H. Metal organic frameworks as multifunctional solid catalysts. *Trends Chem.* **2020**, *2*, 454–466. [CrossRef]
11. Dincă, M.; Gabbaï, F.P.; Long, J.R. Organometallic chemistry within metal–organic frameworks. *Organometallics* **2019**, *38*, 3389–3391. [CrossRef]
12. An, B.; Zeng, L.Z.; Jia, M.; Li, Z.; Lin, Z.K.; Song, Y.; Zhou, Y.; Cheng, J.; Wang, C.; Lin, W.B. Molecular iridium complexes in metal-organic frameworks catalyze CO_2 hydrogenation via concerted proton and hydride transfer. *J. Am. Chem. Soc.* **2017**, *139*, 17747–17750. [CrossRef] [PubMed]
13. Bai, Y.; Dou, Y.; Xie, L.-H.; Rutledge, W.; Li, J.-R.; Zhou, H.-C. Zr-based metal-organic frameworks: Design, synthesis, structure, and applications. *Chem. Soc. Rev.* **2016**, *45*, 2327–2367. [CrossRef] [PubMed]
14. Bosch, M.; Yuan, S.; Rutledge, W.; Zhou, H.-C. Stepwise synthesis of metal-organic frameworks. *Acc. Chem. Res.* **2017**, *50*, 857–865. [CrossRef]
15. Burgess, S.A.; Kassie, A.; Baranowski, S.A.; Fritzsching, K.J.; Schmidt-Rohr, K.; Brown, C.M.; Wade, C.R. Improved catalytic activity and stability of a palladium pincer complex by incorporation into a metal-organic framework. *J. Am. Chem. Soc.* **2016**, *138*, 1780–1783. [CrossRef] [PubMed]
16. Cui, Y.; Li, B.; He, H.; Zhou, W.; Chen, B.; Qian, G. Metal-organic frameworks as platforms for functional materials. *Acc. Chem. Res.* **2016**, *49*, 483–493. [CrossRef]
17. Zhou, H.C.; Long, J.R.; Yaghi, O.M. Introduction to metal-organic frameworks. *Chem. Rev.* **2012**, *112*, 673–674. [CrossRef]

18. Ji, Z.; Wang, H.; Canossa, S.; Wuttke, S.; Yaghi, O.M. Pore chemistry of metal–organic frameworks. *Adv. Funct. Mater.* **2020**, *30*, 2000238. [CrossRef]
19. Kreno, L.E.; Leong, K.; Farha, O.K.; Allendorf, M.; Van Duyne, R.P.; Hupp, J.T. Metal-organic framework materials as chemical sensors. *Chem. Rev.* **2012**, *112*, 1105–1125. [CrossRef] [PubMed]
20. Lee, J.; Farha, O.K.; Roberts, J.; Scheidt, K.A.; Nguyen, S.T.; Hupp, J.T. Metal-organic framework materials as catalysts. *Chem. Soc. Rev.* **2009**, *38*, 1450–1459. [CrossRef]
21. Li, J.R.; Sculley, J.; Zhou, H.C. Metal-organic frameworks for separations. *Chem. Rev.* **2012**, *112*, 869–932. [CrossRef] [PubMed]
22. Nath, I.; Chakraborty, J.; Verpoort, F. Metal organic frameworks mimicking natural enzymes: A structural and functional analogy. *Chem. Soc. Rev.* **2016**, *45*, 4127–4170. [CrossRef]
23. Zhang, Y.M.; Yuan, S.; Day, G.; Wang, X.; Yang, X.Y.; Zhou, H.C. Luminescent sensors based on metal-organic frameworks. *Coord. Chem. Rev.* **2018**, *354*, 28–45. [CrossRef]
24. Furukawa, H.; Cordova, K.E.; O'Keeffe, M.; Yaghi, O.M. The chemistry and applications of metal-organic frameworks. *Science* **2013**, *341*, 974. [CrossRef] [PubMed]
25. Zhang, Y.M.; Yang, X.Y.; Zhou, H.C. Synthesis of MOFs for heterogeneous catalysis via linker design. *Polyhedron* **2018**, *154*, 189–201. [CrossRef]
26. Zhao, M.; Ou, S.; Wu, C.D. Porous Metal-organic frameworks for heterogeneous biomimetic catalysis. *Acc. Chem. Res.* **2014**, *47*, 1199–1207. [CrossRef]
27. Gao, W.Y.; Chrzanowski, M.; Ma, S.Q. Metal-metalloporphyrin frameworks: A resurging class of functional materials. *Chem. Soc. Rev.* **2014**, *43*, 5841–5866. [CrossRef] [PubMed]
28. Feng, D.W.; Chung, W.C.; Wei, Z.W.; Gu, Z.Y.; Jiang, H.L.; Chen, Y.P.; Darensbourg, D.J.; Zhou, H.C. Construction of ultrastable porphyrin zr metal-organic frameworks through linker elimination. *J. Am. Chem. Soc.* **2013**, *135*, 17105–17110. [CrossRef]
29. Huang, N.; Yuan, S.; Drake, H.; Yang, X.Y.; Pang, J.D.; Qin, J.S.; Li, J.L.; Zhang, Y.M.; Wang, Q.; Jiang, D.L.; et al. Systematic engineering of single substitution in zirconium metal-organic frameworks toward high-performance catalysis. *J. Am. Chem. Soc.* **2017**, *139*, 18590–18597. [CrossRef] [PubMed]
30. Fateeva, A.; Chater, P.A.; Ireland, C.P.; Tahir, A.A.; Khimyak, Y.Z.; Wiper, P.V.; Darwent, J.R.; Rosseinsky, M.J. A water-stable porphyrin-based metal-organic framework active for visible-light photocatalysis. *Angew. Chem. Int. Ed.* **2012**, *51*, 7440–7444. [CrossRef]
31. Feng, D.W.; Gu, Z.Y.; Chen, Y.P.; Park, J.; Wei, Z.W.; Sun, Y.J.; Bosch, M.; Yuan, S.; Zhou, H.C. A highly stable porphyrinic zirconium metal-organic framework with shp-a topology. *J. Am. Chem. Soc.* **2014**, *136*, 17714–17717. [CrossRef]
32. Huang, N.; Wang, K.C.; Drake, H.; Cai, P.Y.; Pang, J.D.; Li, J.L.; Che, S.; Huang, L.; Wang, Q.; Zhou, H.C. Tailor-made pyrazolide-based metal-organic frameworks for selective catalysis. *J. Am. Chem. Soc.* **2018**, *140*, 6383–6390. [CrossRef] [PubMed]
33. Wang, C.; Xie, Z.G.; deKrafft, K.E.; Lin, W.L. Doping Metal-organic frameworks for water oxidation, carbon dioxide reduction, and organic photocatalysis. *J. Am. Chem. Soc.* **2011**, *133*, 13445–13454. [CrossRef]
34. Zhu, Y.Y.; Lan, G.X.; Fan, Y.J.; Veroneau, S.S.; Song, Y.; Micheroni, D.; Lin, W.B. Merging photoredox and organometallic catalysts in a metal-organic framework significantly boosts photocatalytic activities. *Angew. Chem. Int. Ed.* **2018**, *57*, 14090–14094. [CrossRef] [PubMed]
35. Lan, G.X.; Li, Z.; Veroneau, S.S.; Zhu, Y.Y.; Xu, Z.W.; Wang, C.; Lin, W.B. Photosensitizing metal-organic layers for efficient sunlight-driven carbon dioxide reduction. *J. Am. Chem. Soc.* **2018**, *140*, 12369–12373. [CrossRef] [PubMed]
36. Elumalai, P.; Mamlouk, H.; Yiming, W.; Feng, L.; Yuan, S.; Zhou, H.C.; Madrahimov, S.T. Recyclable and reusable heteroleptic nickel catalyst immobilized on metal-organic framework for suzuki-miyaura coupling. *ACS Appl. Mater. Interfaces* **2018**, *10*, 41431–41438. [CrossRef] [PubMed]
37. Lv, X.-L.; Wang, K.; Wang, B.; Su, J.; Zou, X.; Xie, Y.; Li, J.-R.; Zhou, H.-C. A base-resistant metalloporphyrin metal-organic framework for C-H bond halogenation. *J. Am. Chem. Soc.* **2017**, *139*, 211–217. [CrossRef]
38. Sikma, R.E.; Kunal, P.; Dunning, S.G.; Reynolds, J.E.; Lee, J.S.; Chang, J.S.; Humphrey, S.M. Organoarsine metal-organic framework with cis-diarsine pockets for the installation of uniquely confined metal complexes. *J. Am. Chem. Soc.* **2018**, *140*, 9806–9809. [CrossRef]
39. Yuan, S.; Qin, J.S.; Su, J.; Li, B.; Li, J.L.; Chen, W.M.; Drake, H.F.; Zhang, P.; Yuan, D.Q.; Zuo, J.L.; et al. Sequential transformation of zirconium(IV)-MOFs into heterobimetallic MOFs bearing magnetic anisotropic cobalt(II) centers. *Angew. Chem. Int. Ed.* **2018**, *57*, 12578–12583. [CrossRef]
40. Yuan, S.; Zhang, P.; Zhang, L.L.; Garcia-Esparza, A.T.; Sokaras, D.; Qin, J.S.; Feng, L.; Day, G.S.; Chen, W.M.; Drake, H.F.; et al. Exposed equatorial positions of metal centers via sequential ligand elimination and installation in MOFs. *J. Am. Chem. Soc.* **2018**, *140*, 10814–10819. [CrossRef]
41. Kosanovich, A.J.; Komatsu, C.H.; Bhuvanesh, N.; Ozerov, O.V.; Perez, L.M. Dearomatization of the PCP pincer ligand in a Re(V) oxo complex. *Chem. Eur. J.* **2018**, *24*, 13754–13757. [CrossRef] [PubMed]
42. He, J.P.; Waggoner, N.W.; Dunning, S.G.; Steiner, A.; Lynch, V.M.; Humphrey, S.M. A PCP Pincer Ligand for coordination polymers with versatile chemical reactivity: Selective activation of CO_2 gas over CO Gas in the solid state. *Angew. Chem. Int. Ed.* **2016**, *55*, 12351–12355. [CrossRef] [PubMed]
43. Carson, F.; Martinez-Castro, E.; Marcos, R.; Miera, G.G.; Jansson, K.; Zou, X.D.; Martin-Matute, B. Effect of the functionalisation route on a Zr-MOF with an Ir-NHC complex for catalysis. *Chem. Commun.* **2015**, *51*, 10864–10867. [CrossRef] [PubMed]

44. Manna, K.; Ji, P.; Lin, Z.; Greene, F.X.; Urban, A.; Thacker, N.C.; Lin, W. Chemoselective single-site Earth-abundant metal catalysts at metal–organic framework nodes. *Nat. Commun.* **2016**, *7*, 12610. [CrossRef] [PubMed]
45. Yuan, S.; Qin, J.S.; Lollar, C.T.; Zhou, H.C. Stable Metal-organic frameworks with group 4 metals: Current status and trends. *ACS Cent. Sci.* **2018**, *4*, 440–450. [CrossRef]
46. Cavka, J.H.; Jakobsen, S.; Olsbye, U.; Guillou, N.; Lamberti, C.; Bordiga, S.; Lillerud, K.P. A new zirconium inorganic building brick forming metal organic frameworks with exceptional stability. *J. Am. Chem. Soc.* **2008**, *130*, 13850–13851. [CrossRef]
47. Schaate, A.; Roy, P.; Godt, A.; Lippke, J.; Waltz, F.; Wiebcke, M.; Behrens, P. Modulated synthesis of Zr-based metal-organic frameworks: From nano to single crystals. *Chem. Eur. J.* **2011**, *17*, 6643–6651. [CrossRef]
48. Yang, D.; Odoh, S.O.; Borycz, J.; Wang, T.C.; Farha, O.K.; Hupp, J.T.; Cramer, C.J.; Gagliardi, L.; Gates, B.C. Tuning Zr_6 metal–organic framework (MOF) nodes as catalyst supports: Site densities and electron-donor properties influence molecular iridium complexes as ethylene conversion catalysts. *ACS Catal.* **2016**, *6*, 235–247. [CrossRef]
49. Denny, M.S.; Parent, L.R.; Patterson, J.P.; Meena, S.K.; Pham, H.; Abellan, P.; Ramasse, Q.M.; Paesani, F.; Gianneschi, N.C.; Cohen, S.M. Transmission electron microscopy reveals deposition of metal oxide coatings onto metal–organic frameworks. *J. Am. Chem. Soc.* **2018**, *140*, 1348–1357. [CrossRef]
50. Liu, L.; Chen, Z.; Wang, J.; Zhang, D.; Zhu, Y.; Ling, S.; Huang, K.-W.; Belmabkhout, Y.; Adil, K.; Zhang, Y.; et al. Imaging defects and their evolution in a metal-organic framework at sub-unit-cell resolution. *Nat. Chem.* **2019**, *11*, 622–628. [CrossRef]
51. Yuan, S.; Zou, L.; Li, H.; Chen, Y.-P.; Qin, J.; Zhang, Q.; Lu, W.; Hall, M.B.; Zhou, H.-C. Flexible zirconium metal-organic frameworks as bioinspired switchable catalysts. *Angew. Chem. Int. Ed.* **2016**, *55*, 10776–10780. [CrossRef]
52. Wang, T.C.; Vermeulen, N.A.; Kim, I.S.; Martinson, A.B.F.; Stoddart, J.F.; Hupp, J.T.; Farha, O.K. Scalable synthesis and post-modification of a mesoporous metal-organic framework called NU-1000. *Nat. Protoc.* **2015**, *11*, 149. [CrossRef]
53. Morris, W.; Volosskiy, B.; Demir, S.; Gandara, F.; McGrier, P.L.; Furukawa, H.; Cascio, D.; Stoddart, J.F.; Yaghi, O.M. Synthesis, structure, and metalation of two new highly porous zirconium metal-organic frameworks. *Inorg. Chem.* **2012**, *51*, 6443–6445. [CrossRef]
54. Liu, T.-F.; Vermeulen, N.A.; Howarth, A.J.; Li, P.; Sarjeant, A.A.; Hupp, J.T.; Farha, O.K. Adding to the arsenal of zirconium-based metal–organic frameworks: The topology as a platform for solvent-assisted metal incorporation. *Eur. J. Inorg. Chem.* **2016**, *2016*, 4349–4352. [CrossRef]
55. Clegg, W.; Harbron, D.R.; Homan, C.D.; Hunt, P.A.; Little, I.R.; Straughan, B.P. Crystal structures of three basic zinc carboxylates together with infrared and FAB mass spectrometry studies in solution. *Inorg. Chim. Acta* **1991**, *186*, 51–60. [CrossRef]
56. Köberl, M.; Cokoja, M.; Herrmann, W.A.; Kühn, F.E. From molecules to materials: Molecular paddle-wheel synthons of macromolecules, cage compounds and metal–organic frameworks. *Dalton Trans.* **2011**, *40*, 6834–6859. [CrossRef] [PubMed]
57. Corma, A.; García, H.; Llabrés i Xamena, F.X. Engineering Metal organic frameworks for heterogeneous catalysis. *Chem. Rev.* **2010**, *110*, 4606–4655. [CrossRef] [PubMed]
58. Yuan, S.; Lu, W.; Chen, Y.-P.; Zhang, Q.; Liu, T.-F.; Feng, D.; Wang, X.; Qin, J.; Zhou, H.-C. Sequential linker installation: Precise placement of functional groups in multivariate metal–organic frameworks. *J. Am. Chem. Soc.* **2015**, *137*, 3177–3180. [CrossRef]
59. Zhang, Y.; Zhang, X.; Lyu, J.; Otake, K.I.; Wang, X.; Redfern, L.R.; Malliakas, C.D.; Li, Z.; Islamoglu, T.; Wang, B.; et al. A flexible metal-organic framework with 4-connected Zr_6 nodes. *J. Am. Chem. Soc.* **2018**, *140*, 11179–11183. [CrossRef] [PubMed]
60. Peters, A.W.; Li, Z.; Farha, O.K.; Hupp, J.T. Toward inexpensive photocatalytic hydrogen evolution: A nickel sulfide catalyst supported on a high-stability metal–organic framework. *ACS Appl. Mater. Interfaces* **2016**, *8*, 20675–20681. [CrossRef] [PubMed]
61. Platero-Prats, A.E.; League, A.B.; Bernales, V.; Ye, J.; Gallington, L.C.; Vjunov, A.; Schweitzer, N.M.; Li, Z.; Zheng, J.; Mehdi, B.L.; et al. Bridging zirconia nodes within a metal–organic framework via catalytic ni-hydroxo clusters to form heterobimetallic nanowires. *J. Am. Chem. Soc.* **2017**, *139*, 10410–10418. [CrossRef]
62. Noh, H.; Cui, Y.; Peters, A.W.; Pahls, D.R.; Ortuño, M.A.; Vermeulen, N.A.; Cramer, C.J.; Gagliardi, L.; Hupp, J.T.; Farha, O.K. An exceptionally stable metal–organic framework supported molybdenum(vi) oxide catalyst for cyclohexene epoxidation. *J. Am. Chem. Soc.* **2016**, *138*, 14720–14726. [CrossRef] [PubMed]
63. Ahn, S.; Thornburg, N.E.; Li, Z.; Wang, T.C.; Gallington, L.C.; Chapman, K.W.; Notestein, J.M.; Hupp, J.T.; Farha, O.K. Stable Metal–organic framework-supported niobium catalysts. *Inorg. Chem.* **2016**, *55*, 11954–11961. [CrossRef]
64. Li, Z.; Peters, A.W.; Bernales, V.; Ortuño, M.A.; Schweitzer, N.M.; DeStefano, M.R.; Gallington, L.C.; Platero-Prats, A.E.; Chapman, K.W.; Cramer, C.J.; et al. Metal–organic framework supported cobalt catalysts for the oxidative dehydrogenation of propane at low temperature. *ACS Cent. Sci.* **2017**, *3*, 31–38. [CrossRef] [PubMed]
65. Noh, H.; Kung, C.-W.; Otake, K.-I.; Peters, A.W.; Li, Z.; Liao, Y.; Gong, X.; Farha, O.K.; Hupp, J.T. Redox-mediator-assisted electrocatalytic hydrogen evolution from water by a molybdenum sulfide-functionalized metal–organic framework. *ACS Catal.* **2018**, *8*, 9848–9858. [CrossRef]
66. Kung, C.-W.; Audu, C.O.; Peters, A.W.; Noh, H.; Farha, O.K.; Hupp, J.T. Copper nanoparticles installed in metal–organic framework thin films are electrocatalytically competent for CO_2 reduction. *ACS Energy Lett.* **2017**, *2*, 2394–2401. [CrossRef]
67. Nguyen, H.G.T.; Schweitzer, N.M.; Chang, C.-Y.; Drake, T.L.; So, M.C.; Stair, P.C.; Farha, O.K.; Hupp, J.T.; Nguyen, S.T. Vanadium-Node-Functionalized UiO-66: A thermally stable MOF-supported catalyst for the gas-phase oxidative dehydrogenation of cyclohexene. *ACS Catal.* **2014**, *4*, 2496–2500. [CrossRef]

68. Yang, D.; Odoh, S.O.; Wang, T.C.; Farha, O.K.; Hupp, J.T.; Cramer, C.J.; Gagliardi, L.; Gates, B.C. Metal–organic framework nodes as nearly ideal supports for molecular catalysts: NU-1000- and UiO-66-supported iridium complexes. *J. Am. Chem. Soc.* **2015**, *137*, 7391–7396. [CrossRef]
69. Pi, Y.; Feng, X.; Song, Y.; Xu, Z.; Li, Z.; Lin, W.B. Metal–organic frameworks integrate cu photosensitizers and secondary building unit-supported fe catalysts for photocatalytic hydrogen evolution. *J. Am. Chem. Soc.* **2020**, *142*, 10302–10307. [CrossRef]
70. An, B.; Li, Z.; Song, Y.; Zhang, J.; Zeng, L.; Wang, C.; Lin, W. Cooperative copper centres in a metal–organic framework for selective conversion of CO_2 to ethanol. *Nat. Chem.* **2019**, *2*, 709–717. [CrossRef]
71. Ji, P.; Manna, K.; Lin, Z.; Feng, X.; Urban, A.; Song, Y.; Lin, W. Single-Site Cobalt Catalysts at New $Zr_{12}(\mu_3\text{-}O)_8(\mu_3\text{-}OH)_8(\mu_2\text{-}OH)_6$ metal–organic framework nodes for highly active hydrogenation of nitroarenes, nitriles, and isocyanides. *J. Am. Chem. Soc.* **2017**, *139*, 7004–7011. [CrossRef] [PubMed]
72. Ji, P.; Song, Y.; Drake, T.; Veroneau, S.S.; Lin, Z.; Pan, X.; Lin, W. Titanium(III)-oxo clusters in a metal–organic framework support single-site Co(II)-hydride catalysts for arene hydrogenation. *J. Am. Chem. Soc.* **2018**, *140*, 433–440. [CrossRef]
73. Comito, R.J.; Fritzsching, K.J.; Sundell, B.J.; Schmidt-Rohr, K.; Dincă, M. Single-site heterogeneous catalysts for olefin polymerization enabled by cation exchange in a metal-organic framework. *J. Am. Chem. Soc.* **2016**, *138*, 10232–10237. [CrossRef]
74. Lollar, C.T.; Qin, J.-S.; Pang, J.; Yuan, S.; Becker, B.; Zhou, H.-C. Interior decoration of stable metal–organic frameworks. *Langmuir* **2018**, *34*, 13795–13807. [CrossRef] [PubMed]
75. Brozek, C.K.; Bellarosa, L.; Soejima, T.; Clark, T.V.; López, N.; Dincă, M. Solvent-dependent cation exchange in metal–organic frameworks. *Chem. Eur. J.* **2014**, *20*, 6871–6874. [CrossRef]
76. Kim, M.; Cahill, J.F.; Fei, H.; Prather, K.A.; Cohen, S.M. Postsynthetic ligand and cation exchange in robust metal–organic frameworks. *J. Am. Chem. Soc.* **2012**, *134*, 18082–18088. [CrossRef] [PubMed]
77. Das, S.; Kim, H.; Kim, K. Metathesis in single crystal: Complete and reversible exchange of metal ions constituting the frameworks of metal−organic frameworks. *J. Am. Chem. Soc.* **2009**, *131*, 3814–3815. [CrossRef] [PubMed]
78. Denysenko, D.; Werner, T.; Grzywa, M.; Puls, A.; Hagen, V.; Eickerling, G.; Jelic, J.; Reuter, K.; Volkmer, D. Reversible gas-phase redox processes catalyzed by Co-exchanged MFU-4*l(arge)*. *Chem. Commun.* **2012**, *48*, 1236–1238. [CrossRef]
79. Sun, D.; Sun, F.; Deng, X.; Li, Z. Mixed-Metal Strategy on Metal–organic frameworks (MOFs) for functionalities expansion: Co substitution induces aerobic oxidation of cyclohexene over inactive Ni-MOF-74. *Inorg. Chem.* **2015**, *54*, 8639–8643. [CrossRef]
80. Zou, L.; Feng, D.; Liu, T.-F.; Chen, Y.-P.; Yuan, S.; Wang, K.; Wang, X.; Fordham, S.; Zhou, H.-C. A versatile synthetic route for the preparation of titanium metal–organic frameworks. *Chem. Sci.* **2016**, *7*, 1063–1069. [CrossRef]
81. Huxley, M.; Coghlan, C.J.; Burgun, A.; Tarzia, A.; Sumida, K.; Sumby, C.J.; Doonan, C.J. Site-specific metal and ligand substitutions in a microporous Mn^{2+}-based metal–organic framework. *Dalton Trans.* **2016**, *45*, 4431–4438. [CrossRef]
82. Banerjee, P.C.; Lobo, D.E.; Middag, R.; Ng, W.K.; Shaibani, M.E.; Majumder, M. Electrochemical capacitance of Ni-doped metal organic framework and reduced graphene oxide composites: More than the sum of its parts. *ACS Appl. Mater. Interfaces* **2015**, *7*, 3655–3664. [CrossRef] [PubMed]
83. Xiao, D.J.; Bloch, E.D.; Mason, J.A.; Queen, W.L.; Hudson, M.R.; Planas, N.; Borycz, J.; Dzubak, A.L.; Verma, P.; Lee, K.; et al. Oxidation of ethane to ethanol by N_2O in a metal–organic framework with coordinatively unsaturated iron(II) sites. *Nat. Chem.* **2014**, *6*, 590. [CrossRef]
84. Li, H.; Shi, W.; Zhao, K.; Li, H.; Bing, Y.; Cheng, P. Enhanced hydrostability in Ni-doped MOF-5. *Inorg. Chem.* **2012**, *51*, 9200–9207. [CrossRef]
85. Botas, J.A.; Calleja, G.; Sánchez-Sánchez, M.; Orcajo, M.G. Cobalt doping of the MOF-5 framework and its effect on gas-adsorption properties. *Langmuir* **2010**, *26*, 5300–5303. [CrossRef] [PubMed]
86. Liu, L.; Telfer, S.G. Systematic ligand modulation enhances the moisture stability and gas sorption characteristics of quaternary metal–organic frameworks. *J. Am. Chem. Soc.* **2015**, *137*, 3901–3909. [CrossRef] [PubMed]
87. Kaye, S.S.; Dailly, A.; Yaghi, O.M.; Long, J.R. Impact of preparation and handling on the hydrogen storage properties of Zn_4O(1,4-benzenedicarboxylate)$_3$ (MOF-5). *J. Am. Chem. Soc.* **2007**, *129*, 14176–14177. [CrossRef] [PubMed]
88. Cychosz, K.A.; Matzger, A.J. Water stability of microporous coordination polymers and the adsorption of pharmaceuticals from water. *Langmuir* **2010**, *26*, 17198–17202. [CrossRef]
89. Drake, T.; Ji, P.; Lin, W. Site isolation in metal-organic frameworks enables novel transition metal catalysis. *Acc. Chem. Res.* **2018**, *51*, 2129–2138. [CrossRef]
90. Pascanu, V.; Miera, G.G.; Inge, A.K.; Martin-Matute, B. Metal–organic frameworks as catalysts for organic synthesis: A critical perspective. *J. Am. Chem. Soc.* **2019**, *141*, 7223–7234. [CrossRef]
91. Paciello, R.A. High Throughput screening of homogeneous catalysts: Selected trends and applications in process development. In *Applied Homogeneous Catalysis with Organometallic Compounds*, 3rd ed.; Wiley-VCH Verlag GmbH & Co. KGaA: Weinheim, Germany, 2019; pp. 1085–1096.
92. Schaub, T.; Paciello, R.A.; Limbach, M. Homogeneous catalysis with CO_2 as a building block: An industrial perspective. In *Applied Homogeneous Catalysis with Organometallic Compounds*, 3rd ed.; Wiley-VCH Verlag GmbH & Co. KGaA: Weinheim, Germany, 2019; pp. 1601–1614.
93. Manna, K.; Ji, P.; Greene, F.X.; Lin, W. Metal-organic framework nodes support single-site magnesium-alkyl catalysts for hydroboration and hydroamination reactions. *J. Am. Chem. Soc.* **2016**, *138*, 7488–7491. [CrossRef]

94. Yang, X.; Yuan, S.; Zou, L.; Drake, H.; Zhang, Y.; Qin, J.; Alsalme, A.; Zhou, H.-C. One-step synthesis of hybrid core-shell metal-organic frameworks. *Angew. Chem. Int. Ed.* **2018**, *57*, 3927–3932. [CrossRef] [PubMed]
95. Thiam, Z.; Abou-Hamad, E.; Dereli, B.; Liu, L.; Emwas, A.-H.; Ahmad, R.; Jiang, H.; Isah, A.A.; Ndiaye, P.B.; Taoufik, M.; et al. Extension of surface organometallic chemistry to metal–organic frameworks: Development of a well-defined single site [(≡Zr–O−)W(=O)(CH_2tBu)$_3$] olefin metathesis catalyst. *J. Am. Chem. Soc.* **2020**, *142*, 16690–16703. [CrossRef]
96. Yuan, S.; Zou, L.; Qin, J.-S.; Li, J.; Huang, L.; Feng, L.; Wang, X.; Bosch, M.; Alsalme, A.; Cagin, T.; et al. Construction of hierarchically porous metal–organic frameworks through linker labilization. *Nat. Commun.* **2017**, *8*, 15356. [CrossRef] [PubMed]
97. Molnar, A. (Ed.) *Palladium-Catalyzed Coupling Reactions: Practical Aspects and Future Developments*; Wiley-VCH Verlag GmbH & Co. KGaA: Weinheim, Germany, 2013; p. 570. [CrossRef]
98. De Vries, A.H.M.; Parlevliet, F.J.; Schmieder-Van De Vondervoort, L.; Mommers, J.H.M.; Henderickx, H.J.W.; Walet, M.A.M.; De Vries, J.G. A practical recycle of a ligand-free palladium catalyst for Heck reactions. *Adv. Synth. Catal.* **2002**, *344*, 996–1002. [CrossRef]
99. Wei, Y.-L.; Li, Y.; Chen, Y.-Q.; Dong, Y.; Yao, J.-J.; Han, X.-Y.; Dong, Y.-B. Pd(II)-NHDC-functionalized UiO-67 type MOF for catalyzing Heck cross-coupling and intermolecular benzyne-benzyne-alkene insertion reactions. *Inorg. Chem.* **2018**, *57*, 4379–4386. [CrossRef]
100. Bahadori, M.; Tangestaninejad, S.; Moghadam, M.; Mirkhani, V.; Mechler, A.; Mohammadpoor-Baltork, I.; Zadehahmadi, F. Metal organic framework-supported N-heterocyclic carbene palladium complex: A highly efficient and reusable heterogeneous catalyst for Suzuki-Miyaura C-C coupling reaction. *Microporous Mesoporous Mater.* **2017**, *253*, 102–111. [CrossRef]
101. Chen, L.; Rangan, S.; Li, J.; Jiang, H.; Li, Y. A molecular Pd(II) complex incorporated into a MOF as a highly active single-site heterogeneous catalyst for C-Cl bond activation. *Green Chem.* **2014**, *16*, 3978–3985. [CrossRef]
102. Fei, H.; Cohen, S.M. A robust, catalytic metal-organic framework with open 2,2′-bipyridine sites. *Chem. Commun.* **2014**, *50*, 4810–4812. [CrossRef]
103. Chen, Y.-Z.; Jiang, H.-L. Porphyrinic metal-organic framework catalyzed Heck-reaction: Fluorescence "turn-on" sensing of Cu(II) ion. *Chem. Mater.* **2016**, *28*, 6698–6704. [CrossRef]
104. Godula, K.; Sames, D. C-H bond functionalization in complex organic synthesis. *Science* **2006**, *312*, 67–72. [CrossRef]
105. Schwach, P.; Pan, X.; Bao, X. Direct conversion of methane to value-added chemicals over heterogeneous catalysts: Challenges and prospects. *Chem. Rev.* **2017**, *117*, 8497–8520. [CrossRef]
106. Liao, P.; Getman, R.B.; Snurr, R.Q. Optimizing open iron sites in metal-organic frameworks for ethane oxidation: A first-principles study. *ACS Appl. Mater. Interfaces* **2017**, *9*, 33484–33492. [CrossRef]
107. Bergman, R.G. Organometallic chemistry: C-H activation. *Nature* **2007**, *446*, 391–393. [CrossRef] [PubMed]
108. Santoro, S.; Kozhushkov, S.I.; Ackermann, L.; Vaccaro, L. Heterogeneous catalytic approaches in C-H activation reactions. *Green Chem.* **2016**, *18*, 3471–3493. [CrossRef]
109. Qin, J.-S.; Yuan, S.; Lollar, C.; Pang, J.; Alsalme, A.; Zhou, H.-C. Stable metal-organic frameworks as a host platform for catalysis and biomimetics. *Chem. Commun.* **2018**, *54*, 4231–4249. [CrossRef]
110. Sawano, T.; Lin, Z.; Boures, D.; An, B.; Wang, C.; Lin, W. Metal-organic frameworks stabilize mono(phosphine)-metal complexes for broad-scope catalytic reactions. *J. Am. Chem. Soc.* **2016**, *138*, 9783–9786. [CrossRef]
111. Zhang, Y.; Li, J.; Yang, X.; Zhang, P.; Pang, J.; Li, B.; Zhou, H.-C. A mesoporous NNN-pincer-based metal-organic framework scaffold for the preparation of noble-metal-free catalysts. *Chem. Commun.* **2019**, *55*, 2023–2026. [CrossRef] [PubMed]
112. Kalaj, M.; Cohen, S.M. Postsynthetic modification: An enabling technology for the advancement of metal–organic frameworks. *ACS Cent. Sci.* **2020**, *6*, 1046–1057. [CrossRef]
113. Anderson, J.S.; Gallagher, A.T.; Mason, J.A.; Harris, T.D. A five-coordinate heme dioxygen adduct isolated within a metal-organic framework. *J. Am. Chem. Soc.* **2014**, *136*, 16489–16492. [CrossRef] [PubMed]
114. Gallagher, A.T.; Lee, J.Y.; Kathiresan, V.; Anderson, J.S.; Hoffman, B.M.; Harris, T.D. A structurally-characterized peroxomanganese(IV) porphyrin from reversible O-2 binding within a metal-organic framework. *Chem. Sci.* **2018**, *9*, 1596–1603. [CrossRef] [PubMed]
115. Brozek, C.K.; Miller, J.T.; Stoian, S.A.; Dincă, M. NO disproportionation at a mononuclear site-isolated Fe^{2+} center in Fe^{2+}-MOF-5. *J. Am. Chem. Soc.* **2015**, *137*, 7495–7501. [CrossRef]
116. Bloch, E.D.; Queen, W.L.; Chavan, S.; Wheatley, P.S.; Zadrozny, J.M.; Morris, R.; Brown, C.M.; Lamberti, C.; Bordiga, S.; Long, J.R. Gradual release of strongly bound nitric oxide from $Fe_2(NO)_2$(dobdc). *J. Am. Chem. Soc.* **2015**, *137*, 3466–3469. [CrossRef]
117. Rieth, A.J.; Tulchinsky, Y.; Dinca, M. High and reversible ammonia uptake in mesoporous azolate metal organic frameworks with open Mn, Co, and Ni sites. *J. Am. Chem. Soc.* **2016**, *138*, 9401–9404. [CrossRef]
118. Reed, D.A.; Keitz, B.K.; Oktawiec, J.; Mason, J.A.; Runcevski, T.; Xiao, D.J.; Darago, L.E.; Crocella, V.; Bordiga, S.; Long, J.R. A spin transition mechanism for cooperative adsorption in metal-organic frameworks. *Nature* **2017**, *550*, 96–100. [CrossRef]
119. Reed, D.A.; Xiao, D.J.; Gonzalez, M.I.; Darago, L.E.; Herm, Z.R.; Grandjean, F.; Long, J.R. Reversible CO scavenging via adsorbate-dependent spin state transitions in an iron(II)-triazolate metal-organic framework. *J. Am. Chem. Soc.* **2016**, *138*, 5594–5602. [CrossRef]
120. Bloch, E.D.; Queen, W.L.; Krishna, R.; Zadrozny, J.M.; Brown, C.M.; Long, J.R. Hydrocarbon separations in a metal-organic framework with open iron(II) coordination sites. *Science* **2012**, *335*, 1606. [CrossRef]

121. Tulchinsky, Y.; Hendon, C.H.; Lomachenko, K.A.; Borfecchia, E.; Melot, B.C.; Hudson, M.R.; Tarver, J.D.; Korzynski, M.D.; Stubbs, A.W.; Kagan, J.J.; et al. Reversible capture and release of Cl_2 and Br_2 with a redox-active metal-organic framework. *J. Am. Chem. Soc.* **2017**, *139*, 5992–5997. [CrossRef] [PubMed]
122. Zeng, M.H.; Yin, Z.; Tan, Y.X.; Zhang, W.X.; He, Y.P.; Kurmoo, M. Nanoporous cobalt(II) MOF exhibiting four magnetic ground states and changes in gas sorption upon post-synthetic modification. *J. Am. Chem. Soc.* **2014**, *136*, 4680–4688. [CrossRef] [PubMed]
123. Motokawa, N.; Miyasaka, H.; Yamashita, M.; Dunbar, K.R. An electron-transfer ferromagnet with T-c=107 K based on a three-dimensional [Ru-2](2)/TCNQ system. *Angew. Chem. Int. Ed.* **2008**, *47*, 7760–7763. [CrossRef]
124. Stone, K.H.; Stephens, P.W.; McConnell, A.C.; Shurdha, E.; Pokhodnya, K.I.; Miller, J.S. Mn-II(TCNE)(3/2)(I-3)(1/2)-A 3D network-structured organic-based magnet and comparison to a 2D analog. *Adv. Mater.* **2010**, *22*, 2514–2519. [CrossRef]
125. Lapidus, S.H.; McConnell, A.C.; Stephens, P.W.; Miller, J.S. Structure and magnetic ordering of a 2-D Mn-II (TCNE)I(OH2) (TCNE = tetracyanoethylene) organic-based magnet (T-c = 171 K). *Chem. Commun.* **2011**, *47*, 7602–7604. [CrossRef] [PubMed]
126. Jeon, I.R.; Negru, B.; Van Duyne, R.P.; Harris, T.D. A 2D Semiquinone Radical-Containing Microporous Magnet with Solvent-Induced Switching from T-c=26 to 80 K. *J. Am. Chem. Soc.* **2015**, *137*, 15699–15702. [CrossRef]
127. DeGayner, J.A.; Jeon, I.R.; Sun, L.; Dinca, M.; Harris, T.D. 2D Conductive Iron-Quinoid Magnets Ordering up to T-c=105 K via Heterogenous Redox Chemistry. *J. Am. Chem. Soc.* **2017**, *139*, 4175–4184. [CrossRef]
128. Zadrozny, J.M.; Gallagher, A.T.; Harris, T.D.; Freedman, D.E. A Porous Array of Clock Qubits. *J. Am. Chem. Soc.* **2017**, *139*, 7089–7094. [CrossRef] [PubMed]
129. Yu, C.-J.; Krzyaniak, M.D.; Fataftah, M.S.; Wasielewski, M.R.; Freedman, D.E. A concentrated array of copper porphyrin candidate qubits. *Chem. Sci.* **2019**. [CrossRef] [PubMed]
130. Taylor-Pashow, K.M.L.; Della Rocca, J.; Xie, Z.G.; Tran, S.; Lin, W.B. Postsynthetic modifications of iron-carboxylate nanoscale metal-organic frameworks for imaging and drug delivery. *J. Am. Chem. Soc.* **2009**, *131*, 14261–14263. [CrossRef] [PubMed]
131. He, C.; Lu, K.; Liu, D.; Lin, W. Nanoscale metal–organic frameworks for the Co-delivery of cisplatin and pooled siRNAs to enhance therapeutic efficacy in drug-resistant ovarian cancer cells. *J. Am. Chem. Soc.* **2014**, *136*, 5181–5184. [CrossRef]
132. Ni, K.Y.; Lan, G.X.; Veroneau, S.S.; Duan, X.P.; Song, Y.; Lin, W.B. Nanoscale metal-organic frameworks for mitochondria-targeted radiotherapy-radiodynamic therapy. *Nat. Commun.* **2018**, *9*, 4321. [CrossRef]
133. Lan, G.X.; Ni, K.Y.; Veroneau, S.S.; Song, Y.; Lin, W.B. Nanoscale metal-organic layers for radiotherapy-radiodynamic therapy. *J. Am. Chem. Soc.* **2018**, *140*, 16971–16975. [CrossRef] [PubMed]
134. Diring, S.; Carne-Sanchez, A.; Zhang, J.C.; Ikemura, S.; Kim, C.; Inaba, H.; Kitagawa, S.; Furukawa, S. Light responsive metal-organic frameworks as controllable CO-releasing cell culture substrates. *Chem. Sci.* **2017**, *8*, 2381–2386. [CrossRef] [PubMed]

inorganics

MDPI

Article

Green Synthesis of a Functionalized Zirconium-Based Metal-Organic Framework for Water and Ethanol Adsorption

Zhijie Chen [1,†], Xingjie Wang [1,2,†], Timur Islamoglu [1,*] and Omar K. Farha [1,*]

1 Department of Chemistry and International Institute of Nanotechnology, Northwestern University, 2145 Sheridan Road, Evanston, IL 60208, USA; zhijie.chen@northwestern.edu (Z.C.); xingjie.wang@northwestern.edu (X.W.)

2 School of Chemistry and Chemical Engineering, South China University of Technology, Guangzhou 510640, China

* Correspondence: timur.islamoglu@northwestern.edu (T.I.); o-farha@northwestern.edu (O.K.F.)

† These authors contributed equally to this work.

Received: 8 April 2019; Accepted: 21 April 2019; Published: 26 April 2019

Abstract: Aqueous synthesis of metal–organic frameworks (MOFs) at room temperature offers many advantages such as reduction in the generation of toxic byproducts and operation costs, as well as increased safety in the material's production. Functional group-bearing MOFs have received growing attention compared to nonfunctionalized analogues due to enhanced adsorption properties of the former in many cases. Here, we report an aqueous solution-based synthesis of a robust zirconium MOF, UiO-66-NO_2, at room temperature. We evaluated the phase purity, porosity, thermal stability, particle morphology and size of the resulting material. High uptake, as well as near complete recyclability of water and ethanol vapor isotherms at room temperature supports the potential of UiO-66-NO_2 as a solid adsorbent in adsorption-based cooling applications or water harvesting systems.

Keywords: zirconium-based MOFs; water adsorption; ethanol adsorption; porous materials; adsorption heat pump

---

1. Introduction

Metal–organic frameworks (MOFs) [1–5] are a promising class of crystalline porous materials with a wide range of applications including but not limited to heterogeneous catalysis [6–10], enzyme and nanoparticle encapsulation [11–14], water capture [15–19], gas storage, and separation [20–25]. Importantly, physical and chemical properties of MOFs can be fine-tuned with the help of reticular chemistry, where pre-selected molecular building blocks are combined to yield pre-designed frameworks for targeted applications [26–30].

While gas adsorption is the most widely studied application of MOFs, vapor adsorption in MOFs has recently garnered some attention especially with the increasing number of water-stable MOFs reported [15–19,31–33]. MOFs with a high capacity for water uptake at room temperature have also gained increasing attention. By tuning features such as the topology, pore size, pore volume, and chemical functionality of MOFs, the relative pressure where the water uptake reaches maximum can be controlled. This is essential for many applications such as natural gas dehydration [34], water capture from air [16,35], adsorption-based cooling systems [15,19,36], and indoor humidity cooling systems [18]. Ethanol can be also used as refrigerant for adsorption-driven cooling application due to its lower freezing point compared to water [37]. Nevertheless, all of these adsorption-based applications such as adsorption beds [21,38], mixed-matrix membranes [39,40], or devices for capturing water from air [35,41], require large amounts of adsorbents for operation. Consequently, developing green and scalable methods for synthesis of MOFs is of interest for the aforementioned applications.

We have recently reported a promising strategy for the scalable synthesis of UiO-66 (UiO = University of Oslo) analogues (i.e., UiO-66-$(COOH)_2$ and UiO-66-F_4) for toxic chemical removal [42]. Herein, we extended this method to the synthesis of a highly stable Zr MOF, UiO-66-NO_2 [43–45]. Various characterization techniques were implemented to confirm phase purity, morphology, thermal stability, and porosity of the resulting MOF. Moreover, we studied water and ethanol vapor adsorption of UiO-66-NO_2, which showed high uptake as well as good recyclability.

2. Results and Discussion

UiO-66 is constructed from 12-connected hexanuclear zirconium oxo-clusters and terephthalate linkers with face-centered cubic (**fcu**) topology [46]. Due to the exceptional chemical and thermal stability, UiO-66 has become one of the most studied MOFs since its invention in 2008 [43,46–49]. UiO-66 can be easily functionalized with different functional groups on account of its isoreticular chemistry [5], however, a vast majority of these studies involve *N*,*N*-dimethylformamide (DMF), a toxic organic solvent, and elevated temperatures [5]. We recently reported an inexpensive, scalable, and environmentally benign procedure to synthesize UiO-66 derivatives (UiO-66-$(COOH)_2$ and UiO-66-F_4) in water at room temperature [42]. Building on the previous successful examples, we extended our strategy to synthesize another UiO-66 derivative, UiO-66-NO_2 (Figure 1A). Among those functionalized UiO-66 analogues, UiO-66-NO_2 is believed to be one of the most chemically stable Zr MOFs which makes it a potential candidate for broader applications [50,51].

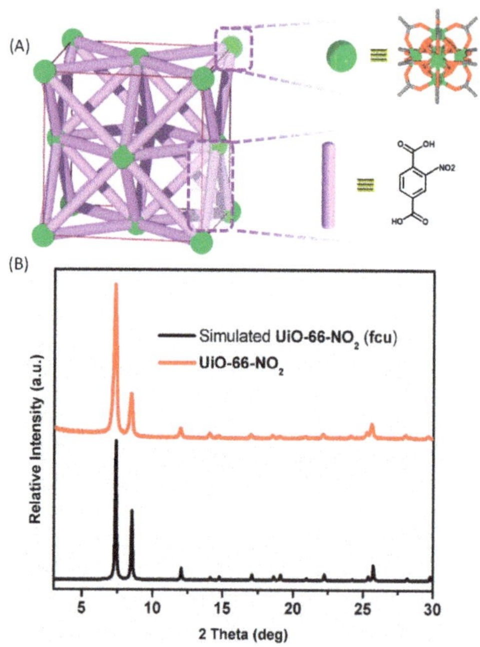

Figure 1. (**A**) Structural representation of UiO-66-NO_2 constructed from 12-c Zr nodes and linear 2-nitroterephthalic acid (H_2BDC-NO_2) ligand. (**B**) Powder X-ray diffraction (PXRD) pattern of the as-synthesized sample matched well with the simulated UiO-66-NO_2.

As illustrated in Figure 1B, our aqueous synthetic method at room temperature resulted in phase-pure UiO-66-NO_2 confirmed by the powder X-ray diffraction (PXRD) pattern. The crystal size of UiO-66-NO_2 was about 100 nm as shown in the scanning electron microscopy (SEM) image (Figure S1).

N_2 adsorption–desorption isotherms at 77 K revealed the experimental total pore volume at $P/P_0 = 0.8$ was about 0.43 $cm^3 \cdot g^{-1}$, and the apparent Brunauer–Emmett–Teller (BET) area of UiO-66-NO_2 was calculated to be 840 $m^2 \cdot g^{-1}$ based on the relative pressure range from $P/P_0 = 0.004$ to $P/P_0 = 0.0294$ (Figure 2A), in line with the surface areas obtained from solvothermal synthesis [52]. The pore-size distribution based on non-local density functional theory (NLDFT) model with the cylindrical kernel showed two major pores. The main peak centered at about 1.3 nm was assigned to octahedral cages of ideal UiO-66 structure, and the smaller shoulder near 1.8 nm was assigned to the presence of non-ordered missing cluster defects (Figure 2B) [53]. The sharp increase in the N_2 isotherm near saturation pressure was due to the condensation of the N_2 gas at interparticle voids, which is common in MOFs with nanosized crystallites [54].

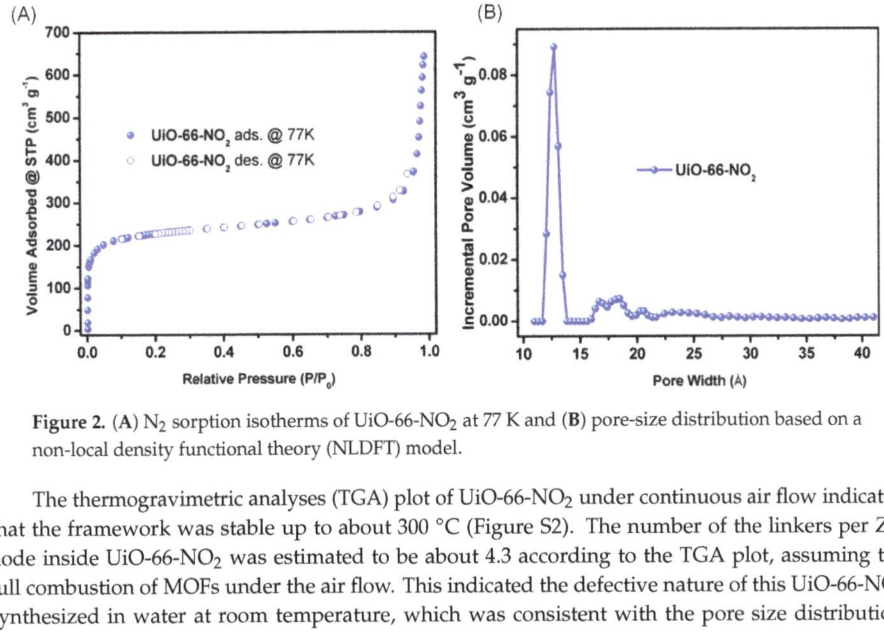

Figure 2. (**A**) N_2 sorption isotherms of UiO-66-NO_2 at 77 K and (**B**) pore-size distribution based on a non-local density functional theory (NLDFT) model.

The thermogravimetric analyses (TGA) plot of UiO-66-NO_2 under continuous air flow indicated that the framework was stable up to about 300 °C (Figure S2). The number of the linkers per Zr_6 node inside UiO-66-NO_2 was estimated to be about 4.3 according to the TGA plot, assuming the full combustion of MOFs under the air flow. This indicated the defective nature of this UiO-66-NO_2 synthesized in water at room temperature, which was consistent with the pore size distribution, further indicating the presence of defects. Diffuse reflectance infrared Fourier transform spectroscopy (DRIFTS) spectra was collected on this material where a sharp band at 3667 cm^{-1} was assigned to μ_3 -OH stretches on the hexanuclear zirconium oxo-clusters (Figure S3) which was red-shifted compared to its parent UiO-66 [55] and consistent with solvothermally synthesized UiO-66-NO_2 [52]

Encouraged by the moderate surface area and high stability of the UiO-66-NO_2, we further studied the ethanol and water vapor sorption, which is important for practical applications such as adsorption-based heat pump and vapor capture from air, respectively. Ethanol adsorption of UiO-66-NO_2 at 298 K showed an uptake of about 112 $cm^3 \cdot g^{-1}$ at $P/P_0 = 0.3$ and about 186 $cm^3 \cdot g^{-1}$ at $P/P_0 = 0.9$, respectively (Figure 3). Moreover, it retained its ethanol sorption performance for the second cycle after the simple activation under dynamic vacuum.

Water adsorption isotherm at room temperature is important to gauge MOFs for their potential for water vapor sorption related applications or to assess their moisture stability [19,35]. Thus, we performed the water vapor sorption at 298 K for UiO-66-NO_2, which exhibited an S-shape adsorption isotherm (Figure 4). The uptake of water vapor at $P/P_0 = 0.9$ was about 459 $cm^3 \cdot g^{-1}$ (or 0.37 $g \cdot g^{-1}$), and this value was slightly smaller than the reported value [17] of parent UiO-66 (533 $cm^3 \cdot g^{-1}$ or

0.43 g·g^{-1}), due to the reduction of pore volume after the introduction of nitro-groups. On the other hand, the water uptake of UiO-66-NO_2 at P/P_0 = 0.3 was about 177 cm^3·g^{-1} (or 0.14 g·g^{-1}) which was higher than that of UiO-66 (108 cm^3·g^{-1} or 0.09 g·g^{-1}), indicating the more hydrophilic nature of this material in comparison with the parent UiO-66. The second run of the same batch material also verified the good cycling ability and therefore the stability of UiO-66-NO_2. Note, the UiO-66-NO_2 synthesized in aqueous solution at room temperature showed similar water uptake to that previously reported (459 cm^3·g^{-1} (or 20.5 mmol·g^{-1}) in this work, in comparison to 533 cm^3·g^{-1} (or 22 mmol·g^{-1}) at P/P_0 = 0.9 by Walton and coworkers [44]), further confirming the practicality of this synthesis method of UiO-66-NO_2 compared to that of solvothermal based methods.

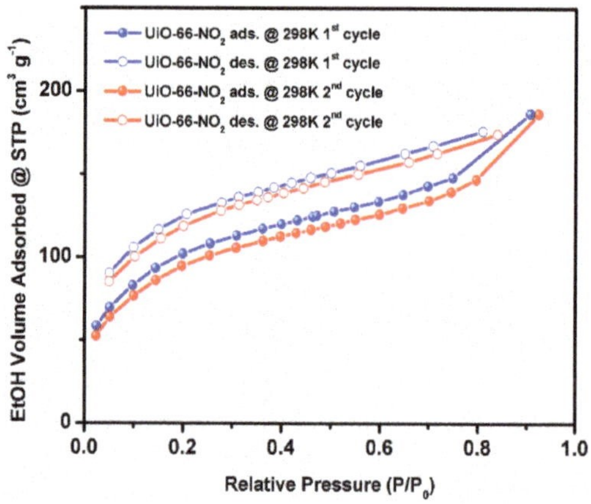

Figure 3. Ethanol sorption isotherms of UiO-66-NO_2 at 298 K (two cycles).

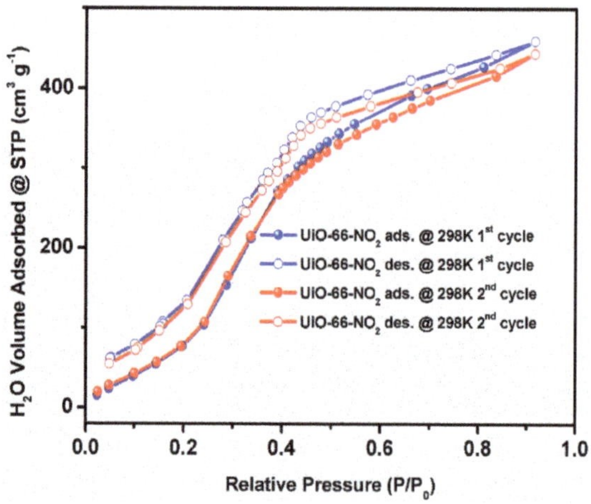

Figure 4. Water sorption isotherms of UiO-66-NO_2 at 298 K (two cycles).

3. Materials and Methods

All reagents were obtained from commercial sources and used without further purification, unless otherwise noted. Zirconium (IV) oxynitrate hydrate (99%) was purchased from Sigma-Aldrich (St. Louis, MO, USA).

3.1. X-ray Diffraction Analyses

Powder X-ray diffraction (PXRD) of MOFs were measured at room temperature on a STOE-STADIMP powder diffractometer (STOE & Cie GmbH, Darmstadt, Germany) equipped with an asymmetric curved Germanium monochromator (Cu Kα1 radiation, λ = 1.54056 Å) and one-dimensional silicon strip detector (MYTHEN2 1K from DECTRIS, Baden, Switzerland). The line focused Cu X-ray tube was operated at 40 kV and 40 mA. The activated powder was sandwiched between two Kapton foils and measured in transmission geometry in a rotating holder. Intensity data from 1 to 30 degrees two theta were collected over a period of 6 min. The instrument was calibrated against a NIST Silicon standard (640d) prior to the measurement.

3.2. N_2 Sorption Measurements

N_2 adsorption and desorption isotherms on activated materials were measured on a Tristar (Micromeritics, Norcross, GA, USA) instrument at 77 K. In general, about 30–50 mg of sample was used in each measurement after activated at 120 °C for 24 h.

3.3. Thermogravimetric Analyses (TGA)

TGA was performed on a TGA/DCS 1 system (Mettler-Toledo AG, Schwerzenbach, Switzerland), which runs on a PC with STARe software. Samples were heated from 30 to 600 °C at a rate of 10 °C/min under air with flow rate 20 mL/min.

3.4. Diffuse Reflectance for Infrared Fourier Transform Spectroscopy (DRIFTS)

DRIFTS spectra were recorded on a Nicolet 6700 FTIR spectrometer (Thermo Nicolet Corp., Madison, WI, USA) equipped with an MCT detector. The detector was cooled with liquid N_2 and the spectra were collected under Ar atmosphere. KBr was utilized as a background spectrum.

3.5. Scanning Electron Micrographs (SEM)

Scanning electron micrographs (SEM) images were taken using a Hitachi SU8030 (Hitachi High Technologies Corporation, Tokyo, Japan) at the EPIC facility (NUANCE Center-Northwestern University). Samples were activated and coated with OsO_4 to ~9 nm thickness in a Denton Desk III TSC Sputter Coater before imaging.

3.6. Water and Ethanol Vapor Sorption Measurements

Water and ethanol isotherms were measured on a 3Flex (Micromeritics), and the water uptake in g·g^{-1} unit is calculated as [(adsorbed amount of water)/(amount of adsorbent)]. Prior to the adsorption measurements, water or ethanol (analyte) was flash frozen under liquid nitrogen and then evacuated under dynamic vacuum at least 3 times to remove any gases in the water reservoir until the P0 of the analyte gas matches the suggested P0 at room temperature. The measurement temperature was controlled with an ISO Controller (Micromeritics).

3.7. Synthesis of UiO-66-NO_2

1.5 mL of water was added to 20 mg (~0.094 mmol) of 2-nitroterephthalic acid. The mixture was then placed in the oven at 100 °C for 5 min. The color of the solution became pale yellow. Separately, 48 mg (~0.208 mmol) of zirconyl nitrate hydrate ($ZrO(NO_3)_2{\cdot}xH_2O$) was dissolved by adding 1.0 mL of water followed by sonication. The two solutions were combined, followed by an addition of 300 µL (~5.24 mmol) of acetic acid. The mixture was stirred for 72 h at room temperature. The crystalline powders were collected by centrifugation (7000 rpm for 30 min). As-synthesized sample was washed 3 times with DI water and then sequentially washed 2 times with ethanol, and finally immersed in acetone for 2 days, during which time the acetone was replaced 2 times per day. The product was obtained after initial drying via leaving exposed to ambient atmosphere overnight and activated at 120 °C for 24 h; yield: 25 mg.

4. Conclusions

In conclusion, we extended the room temperature, aqueous methodology of Zr MOFs to access a nitro-functionalized UiO-66 derivative, UiO-66-NO_2. Importantly, the UiO-66-NO_2 obtained here, using more environmentally benign methods, showed comparable BET area to MOFs obtained from solvothermal synthesis. Moreover, water and ethanol vapor sorption isotherms at room temperature indicated high uptake of both vapors, which suggests potential of UiO-66-NO_2 for adsorption-based cooling applications or water harvesting systems.

Supplementary Materials: The following are available online at http://www.mdpi.com/2304-6740/7/5/56/s1, Figures S1–S3: SEM image, TGA curve and DRIFTS spectra of UiO-66-NO_2.

Author Contributions: Conceptualization, Z.C., T.I. and O.K.F.; methodology, Z.C.; software, Z.C.; validation, T.I. and O.K.F.; formal analysis, X.W., Z.C.; investigation, Z.C. and X.W.; resources, O.K.F.; data curation, Z.C. and X.W.; writing—original draft preparation, Z.C.; writing—review and editing, X.W., T.I. and O.K.F.; visualization, Z.C. and X.W.; supervision, T.I. and O.K.F.; project administration, Z.C., T.I. and O.K.F.; funding acquisition, O.K.F.

Funding: This research was funded by STIR from ARO, grant number W911NF-18-1-0050.

Acknowledgments: The authors thank Zoha Syed for helpful discussions. This work made use of the EPIC facility of Northwestern University's NUANCE Center, which has received support from the Soft and Hybrid Nanotechnology Experimental (SHyNE) Resource (NSF NNCI-1542205); the MRSEC program (NSF DMR-1720139) at the Materials Research Center; the International Institute for Nanotechnology (IIN); the Keck Foundation; and the State of Illinois, through the IIN. This work made use of the IMSERC at Northwestern University, which has received support from the NSF (CHE-1048773 and DMR0521267); Soft and Hybrid Nanotechnology Experimental (SHyNE) Resource (NSF NNCI-1542205); the State of Illinois and International Institute for Nanotechnology (IIN). X.W. acknowledges support from China Scholarship Council (CSC) during his visit to Northwestern University.

Conflicts of Interest: The authors declare no conflict of interest.

References

1. Furukawa, H.; Cordova, K.E.; O'Keeffe, M.; Yaghi, O.M. The Chemistry and Applications of Metal-Organic Frameworks. *Science* **2013**, *341*, 1230444.
2. Férey, G. Building Units Design and Scale Chemistry. *J. Solid State Chem.* **2000**, *152*, 37–48. [CrossRef]
3. Yuan, S.; Feng, L.; Wang, K.; Pang, J.; Bosch, M.; Lollar, C.; Sun, Y.; Qin, J.; Yang, X.; Zhang, P.; et al. Stable Metal-Organic Frameworks: Design, Synthesis, and Applications. *Adv. Mater.* **2018**, *30*, 1704303.
4. Horike, S.; Shimomura, S.; Kitagawa, S. Soft porous crystals. *Nat. Chem.* **2009**, *1*, 695–704. [CrossRef] [PubMed]
5. Chen, Z.; Hanna, S.L.; Redfern, L.R.; Alezi, D.; Islamoglu, T.; Farha, O.K. Reticular chemistry in the rational synthesis of functional zirconium cluster-based MOFs. *Coord. Chem. Rev.* **2019**, *386*, 32–49. [CrossRef]
6. Zhang, X.; Huang, Z.; Ferrandon, M.; Yang, D.; Robison, L.; Li, P.; Wang, T.C.; Delferro, M.; Farha, O.K. Catalytic chemoselective functionalization of methane in a metal–organic framework. *Nat. Catal.* **2018**, *1*, 356–362.
7. Zhang, X.; Vermeulen, N.A.; Huang, Z.; Cui, Y.; Liu, J.; Krzyaniak, M.D.; Li, Z.; Noh, H.; Wasielewski, M.R.; Delferro, M.; et al. Effect of Redox "Non-Innocent" Linker on the Catalytic Activity of Copper-Catecholate-Decorated Metal-Organic Frameworks. *ACS Appl. Mater. Interfaces* **2018**, *10*, 635–641. [CrossRef] [PubMed]
8. Liu, J.; Ye, J.; Li, Z.; Otake, K.-i.; Liao, Y.; Peters, A.W.; Noh, H.; Truhlar, D.G.; Gagliardi, L.; Cramer, C.J.; et al. Beyond the Active Site: Tuning the Activity and Selectivity of a Metal-Organic Framework-Supported Ni Catalyst for Ethylene Dimerization. *J. Am. Chem. Soc.* **2018**, *140*, 11174–11178. [CrossRef] [PubMed]
9. Chen, X.; Jiang, H.; Hou, B.; Gong, W.; Liu, Y.; Cui, Y. Boosting Chemical Stability, Catalytic Activity, and Enantioselectivity of Metal-Organic Frameworks for Batch and Flow Reactions. *J. Am. Chem. Soc.* **2017**, *139*, 13476–13482. [CrossRef] [PubMed]
10. Chen, Z.; Islamoglu, T.; Farha, O.K. Toward Base Heterogenization: A Zirconium Metal-Organic Framework/Dendrimer or Polymer Mixture for Rapid Hydrolysis of a Nerve-Agent Simulant. *ACS Appl. Nano Mater.* **2019**, *2*, 1005–1008. [CrossRef]

11. Wang, S.; Chen, Y.; Wang, S.; Li, P.; Mirkin, C.A.; Farha, O.K. DNA-Functionalized Metal-Organic Framework Nanoparticles for Intracellular Delivery of Proteins. *J. Am. Chem. Soc.* **2019**, *141*, 2215–2219. [CrossRef]
12. Lian, X.; Fang, Y.; Joseph, E.; Wang, Q.; Li, J.; Banerjee, S.; Lollar, C.; Wang, X.; Zhou, H.-C. Enzyme–MOF (metal–organic framework) composites. *Chem. Soc. Rev.* **2017**, *46*, 3386–3401. [CrossRef]
13. Li, G.; Zhao, S.; Zhang, Y.; Tang, Z. Metal-Organic Frameworks Encapsulating Active Nanoparticles as Emerging Composites for Catalysis: Recent Progress and Perspectives. *Adv. Mater.* **2018**, *30*, 1800702.
14. Osterrieth, J.W.M.; Wright, D.; Noh, H.; Kung, C.-W.; Vulpe, D.; Li, A.; Park, J.E.; Van Duyne, R.P.; Moghadam, P.Z.; Baumberg, J.J.; et al. Core–Shell Gold Nanorod@Zirconium-Based Metal-Organic Framework Composites as in Situ Size-Selective Raman Probes. *J. Am. Chem. Soc.* **2019**, *141*, 3893–3900. [CrossRef]
15. Rieth, A.J.; Yang, S.; Wang, E.N.; Dincă, M. Record Atmospheric Fresh Water Capture and Heat Transfer with a Material Operating at the Water Uptake Reversibility Limit. *ACS Cent. Sci.* **2017**, *3*, 668–672. [CrossRef]
16. Kim, H.; Yang, S.; Rao, S.R.; Narayanan, S.; Kapustin, E.A.; Furukawa, H.; Umans, A.S.; Yaghi, O.M.; Wang, E.N. Water harvesting from air with metal–organic frameworks powered by natural sunlight. *Science* **2017**, *356*, 430–434. [CrossRef] [PubMed]
17. Furukawa, H.; Gándara, F.; Zhang, Y.-B.; Jiang, J.; Queen, W.L.; Hudson, M.R.; Yaghi, O.M. Water Adsorption in Porous Metal-Organic Frameworks and Related Materials. *J. Am. Chem. Soc.* **2014**, *136*, 4369–4381. [CrossRef]
18. Towsif Abtab, S.M.; Alezi, D.; Bhatt, P.M.; Shkurenko, A.; Belmabkhout, Y.; Aggarwal, H.; Weseliński, Ł.J.; Alsadun, N.; Samin, U.; Hedhili, M.N.; et al. Reticular Chemistry in Action: A Hydrolytically Stable MOF Capturing Twice Its Weight in Adsorbed Water. *Chem* **2018**, *4*, 94–105. [CrossRef]
19. Chen, Z.; Li, P.; Zhang, X.; Li, P.; Wasson, M.C.; Islamoglu, T.; Stoddart, J.F.; Farha, O.K. Reticular Access to Highly Porous acs-MOFs with Rigid Trigonal Prismatic Linkers for Water Sorption. *J. Am. Chem. Soc.* **2019**, *141*, 2900–2905. [CrossRef]
20. Cadiau, A.; Adil, K.; Bhatt, P.M.; Belmabkhout, Y.; Eddaoudi, M. A metal–organic framework-based splitter for separating propylene from propane. *Science* **2016**, *353*, 137–140. [CrossRef]
21. Lin, R.-B.; Li, L.; Zhou, H.-L.; Wu, H.; He, C.; Li, S.; Krishna, R.; Li, J.; Zhou, W.; Chen, B. Molecular sieving of ethylene from ethane using a rigid metal–organic framework. *Nat. Mater.* **2018**, *17*, 1128–1133. [CrossRef] [PubMed]
22. Cui, X.; Chen, K.; Xing, H.; Yang, Q.; Krishna, R.; Bao, Z.; Wu, H.; Zhou, W.; Dong, X.; Han, Y.; et al. Pore chemistry and size control in hybrid porous materials for acetylene capture from ethylene. *Science* **2016**, *353*, 141–144. [CrossRef] [PubMed]
23. Chen, Z.; Adil, K.; Weselinski, L.J.; Belmabkhout, Y.; Eddaoudi, M. A supermolecular building layer approach for gas separation and storage applications: The eea and rtl MOF platforms for CO_2 capture and hydrocarbon separation. *J. Mater. Chem. A* **2015**, *3*, 6276–6281. [CrossRef]
24. Chen, Z.; Feng, L.; Liu, L.; Bhatt, P.M.; Adil, K.; Emwas, A.-H.; Assen, A.H.; Belmabkhout, Y.; Han, Y.; Eddaoudi, M. Enhanced Separation of Butane Isomers via Defect Control in a Fumarate/Zirconium-Based Metal Organic Framework. *Langmuir* **2018**, *34*, 14546–14551. [CrossRef] [PubMed]
25. Belmabkhout, Y.; Pillai, R.S.; Alezi, D.; Shekhah, O.; Bhatt, P.M.; Chen, Z.; Adil, K.; Vaesen, S.; De Weireld, G.; Pang, M.; et al. Metal–organic frameworks to satisfy gas upgrading demands: Fine-tuning the soc-MOF platform for the operative removal of H2S. *J. Mater. Chem. A* **2017**, *5*, 3293–3303. [CrossRef]
26. Yaghi, O.M.; O'Keeffe, M.; Ockwig, N.W.; Chae, H.K.; Eddaoudi, M.; Kim, J. Reticular synthesis and the design of new materials. *Nature* **2003**, *423*, 705–714. [CrossRef]
27. Rungtaweevoranit, B.; Diercks, C.S.; Kalmutzki, M.J.; Yaghi, O.M. Spiers Memorial Lecture: Progress and prospects of reticular chemistry. *Faraday Discuss.* **2017**, *201*, 9–45.
28. Chen, Z.; Weseliński, Ł.J.; Adil, K.; Belmabkhout, Y.; Shkurenko, A.; Jiang, H.; Bhatt, P.M.; Guillerm, V.; Dauzon, E.; Xue, D.-X.; O'Keeffe, M.; et al. Applying the Power of Reticular Chemistry to Finding the Missing alb-MOF Platform Based on the (6,12)-Coordinated Edge-Transitive Net. *J. Am. Chem. Soc.* **2017**, *139*, 3265–3274. [CrossRef]
29. Chen, Z.; Jiang, H.; O'Keeffe, M.; Eddaoudi, M. Minimal edge-transitive nets for the design and construction of metal–organic frameworks. *Faraday Discuss.* **2017**, *201*, 127–143. [CrossRef]
30. Jiang, H.; Jia, J.; Shkurenko, A.; Chen, Z.; Adil, K.; Belmabkhout, Y.; Weselinski, L.J.; Assen, A.H.; Xue, D.-X.; O'Keeffe, M.; et al. Enriching the Reticular Chemistry Repertoire: Merged Nets Approach for the Rational Design of Intricate Mixed-Linker Metal-Organic Framework Platforms. *J. Am. Chem. Soc.* **2018**, *140*, 8858–8867. [CrossRef]

31. Jeremias, F.; Fröhlich, D.; Janiak, C.; Henninger, S.K. Water and methanol adsorption on MOFs for cycling heat transformation processes. *New J. Chem.* **2014**, *38*, 1846–1852. [CrossRef]
32. Wang, S.; Lee, J.S.; Wahiduzzaman, M.; Park, J.; Muschi, M.; Martineau-Corcos, C.; Tissot, A.; Cho, K.H.; Marrot, J.; Shepard, W.; et al. A robust large-pore zirconium carboxylate metal–organic framework for energy-efficient water-sorption-driven refrigeration. *Nat. Energy* **2018**, *3*, 985–993. [CrossRef]
33. Akiyama, G.; Matsuda, R.; Sato, H.; Hori, A.; Takata, M.; Kitagawa, S. Effect of functional groups in MIL-101 on water sorption behavior. *Microporous Mesoporous Mater.* **2012**, *157*, 89–93. [CrossRef]
34. Cadiau, A.; Belmabkhout, Y.; Adil, K.; Bhatt, P.M.; Pillai, R.S.; Shkurenko, A.; Martineau-Corcos, C.; Maurin, G.; Eddaoudi, M. Hydrolytically stable fluorinated metal–organic frameworks for energy-efficient dehydration. *Science* **2017**, *356*, 731–735. [CrossRef]
35. Fathieh, F.; Kalmutzki, M.J.; Kapustin, E.A.; Waller, P.J.; Yang, J.; Yaghi, O.M. Practical water production from desert air. *Sci. Adv.* **2018**, *4*, eaat3198. [CrossRef]
36. Chaemchuen, S.; Xiao, X.; Klomkliang, N.; Yusubov, M.S.; Verpoort, F. Tunable Metal-Organic Frameworks for Heat Transformation Applications. *Nanomaterials* **2018**, *8*, 661. [CrossRef]
37. De Lange, M.F.; van Velzen, B.L.; Ottevanger, C.P.; Verouden, K.J.F.M.; Lin, L.-C.; Vlugt, T.J.H.; Gascon, J.; Kapteijn, F. Metal-Organic Frameworks in Adsorption-Driven Heat Pumps: The Potential of Alcohols as Working Fluids. *Langmuir* **2015**, *31*, 12783–12796. [CrossRef]
38. Liao, P.-Q.; Huang, N.-Y.; Zhang, W.-X.; Zhang, J.-P.; Chen, X.-M. Controlling guest conformation for efficient purification of butadiene. *Science* **2017**, *356*, 1193. [CrossRef]
39. Lin, R.; Villacorta Hernandez, B.; Ge, L.; Zhu, Z. Metal organic framework based mixed matrix membranes: An overview on filler/polymer interfaces. *J. Mater. Chem. A* **2018**, *6*, 293–312. [CrossRef]
40. Liu, Y.; Chen, Z.; Liu, G.; Belmabkhout, Y.; Adil, K.; Eddaoudi, M.; Koros, W. Conformation-Controlled Molecular Sieving Effects for Membrane-Based Propylene/Propane Separation. *Adv. Mater.* **2019**, *0*, 1807513. [CrossRef]
41. Kalmutzki, M.J.; Diercks, C.S.; Yaghi, O.M. Metal-Organic Frameworks for Water Harvesting from Air. *Adv. Mater.* **2018**, *30*, e1704304. [CrossRef] [PubMed]
42. Chen, Z.; Wang, X.; Noh, H.; Ayoub, G.; Peterson, G.W.; Buru, C.T.; Islamoglu, T.; Farha, O.K. Scalable, room temperature, and water-based synthesis of functionalized zirconium-based metal–organic frameworks for toxic chemical removal. *CrystEngComm* **2019**, *21*, 2409–2415. [CrossRef]
43. Garibay, S.J.; Cohen, S.M. Isoreticular synthesis and modification of frameworks with the UiO-66 topology. *Chem. Commun.* **2010**, *46*, 7700–7702. [CrossRef] [PubMed]
44. Cmarik, G.E.; Kim, M.; Cohen, S.M.; Walton, K.S. Tuning the Adsorption Properties of UiO-66 via Ligand Functionalization. *Langmuir* **2012**, *28*, 15606–15613. [CrossRef]
45. Rada, Z.H.; Abid, H.R.; Shang, J.; Sun, H.; He, Y.; Webley, P.; Liu, S.; Wang, S. Functionalized UiO-66 by Single and Binary $(OH)_2$ and NO_2 Groups for Uptake of CO_2 and CH_4. *Ind. Eng. Chem. Res.* **2016**, *55*, 7924–7932. [CrossRef]
46. Cavka, J.H.; Jakobsen, S.; Olsbye, U.; Guillou, N.; Lamberti, C.; Bordiga, S.; Lillerud, K.P. A New Zirconium Inorganic Building Brick Forming Metal Organic Frameworks with Exceptional Stability. *J. Am. Chem. Soc.* **2008**, *130*, 13850–13851. [CrossRef]
47. DeStefano, M.R.; Islamoglu, T.; Garibay, S.J.; Hupp, J.T.; Farha, O.K. Room-Temperature Synthesis of UiO-66 and Thermal Modulation of Densities of Defect Sites. *Chem. Mater.* **2017**, *29*, 1357–1361. [CrossRef]
48. Katz, M.J.; Mondloch, J.E.; Totten, R.K.; Park, J.K.; Nguyen, S.T.; Farha, O.K.; Hupp, J.T. Simple and Compelling Biomimetic Metal-Organic Framework Catalyst for the Degradation of Nerve Agent Simulants. *Angew. Chem. Int. Ed.* **2014**, *53*, 497–501. [CrossRef]
49. Taddei, M.; Wakeham, R.J.; Koutsianos, A.; Andreoli, E.; Barron, A.R. Post-Synthetic Ligand Exchange in Zirconium-Based Metal-Organic Frameworks: Beware of The Defects! *Angew. Chem. Int. Ed.* **2018**, *57*, 11706–11710. [CrossRef]
50. Kandiah, M.; Nilsen, M.H.; Usseglio, S.; Jakobsen, S.; Olsbye, U.; Tilset, M.; Larabi, C.; Quadrelli, E.A.; Bonino, F.; Lillerud, K.P. Synthesis and Stability of Tagged UiO-66 Zr-MOFs. *Chem. Mater.* **2010**, *22*, 6632–6640. [CrossRef]
51. Howarth, A.J.; Liu, Y.; Li, P.; Li, Z.; Wang, T.C.; Hupp, J.T.; Farha, O.K. Chemical, thermal and mechanical stabilities of metal–organic frameworks. *Nat. Rev. Mater.* **2016**, *1*, 15018. [CrossRef]

52. Wei, R.; Gaggioli, C.A.; Li, G.; Islamoglu, T.; Zhang, Z.; Yu, P.; Farha, O.K.; Cramer, C.J.; Gagliardi, L.; Yang, D.; et al. Tuning the Properties of Zr_6O_8 Nodes in the Metal Organic Framework UiO-66 by Selection of Node-Bound Ligands and Linkers. *Chem. Mater.* **2019**, *31*, 1655–1663. [CrossRef]
53. Shearer, G.C.; Vitillo, J.G.; Bordiga, S.; Svelle, S.; Olsbye, U.; Lillerud, K.P. Functionalizing the Defects: Postsynthetic Ligand Exchange in the Metal Organic Framework UiO-66. *Chem. Mater.* **2016**, *28*, 7190–7193. [CrossRef]
54. Majewski, M.B.; Noh, H.; Islamoglu, T.; Farha, O.K. NanoMOFs: Little crystallites for substantial applications. *J. Mater. Chem. A* **2018**, *6*, 7338–7350. [CrossRef]
55. Islamoglu, T.; Ray, D.; Li, P.; Majewski, M.B.; Akpinar, I.; Zhang, X.; Cramer, C.J.; Gagliardi, L.; Farha, O.K. From Transition Metals to Lanthanides to Actinides: Metal-Mediated Tuning of Electronic Properties of Isostructural Metal-Organic Frameworks. *Inorg. Chem.* **2018**, *57*, 13246–13251. [CrossRef]

inorganics

MDPI

Article

Post Synthetic Defect Engineering of UiO-66 Metal–Organic Framework with An Iridium(III)-HEDTA Complex and Application in Water Oxidation Catalysis

Giordano Gatto, Alceo Macchioni *, Roberto Bondi, Fabio Marmottini and Ferdinando Costantino *

Department of Chemistry, Biology and Biotechnology, Università di Perugia and CIRCC, Via Elce di Sotto, 8, I-06123 Perugia, Italy; giordano.gatto@libero.it (G.G.); robertobondi11@gmail.com (R.B.); fabio.marmottini@unipg.it (F.M.)

* Correspondence: alceo.macchioni@unipg.it (A.M.); ferdinando.costantino@unipg.it (F.C.); Tel.: +39-075-585-5579 (A.M.); +39-075-585-5563 (F.C.)

Received: 9 August 2019; Accepted: 5 October 2019; Published: 10 October 2019

Abstract: Clean production of renewable fuels is a great challenge of our scientific community. Iridium complexes have demonstrated a superior catalytic activity in the water oxidation (WO) reaction, which is a crucial step in water splitting process. Herein, we have used a defective zirconium metal–organic framework (MOF) with UiO-66 structure as support of a highly active Ir complex based on EDTA with the formula [Ir(HEDTA)Cl]Na. The defects are induced by the partial substitution of terephthalic acid with smaller formate groups. Anchoring of the complex occurs through a post-synthetic exchange of formate anions, coordinated at the zirconium clusters of the MOF, with the free carboxylate group of the $[Ir(HEDTA)Cl]^-$ complex. The modified material was tested as a heterogeneous catalyst for the WO reaction by using cerium ammonium nitrate (CAN) as the sacrificial agent. Although turnover frequency (TOF) and turnover number (TON) values are comparable to those of other iridium heterogenized catalysts, the MOF exhibits iridium leaching not limited at the first catalytic run, as usually observed, suggesting a lack of stability of the hybrid system under strong oxidative conditions.

Keywords: metal–organic framework; post-synthetic modification; iridium catalysis; water oxidation; water splitting

1. Introduction

Water oxidation (WO) to molecular oxygen is considered the ideal reaction to provide electrons and protons for the generation of renewable fuels [1–3]. In addition to being thermodynamically disfavored, WO is also an extremely complicated, multi-electron and multi-proton reaction from the kinetic point of view, asking for an efficient and robust catalyst [4]. WOCs (water oxidation catalysts) based on iridium are among the most efficient reported in the literature so far, having, however, in the little abundance and, consequently, high cost of iridium their "Achilles' heel" [5–7]. A possible strategy to alleviate this problem stems in the minimization of the amount of noble-metal exploited in the catalytic process, according to the noble-metal atom economy principle [5]. This can be accomplished by utilizing (i) extremely active molecular catalysts at very low concentration [8–15], (ii) layered heterogeneous catalyst in which almost all active sites are reachable by the substrate [16], and (iii) heterogenized hybrid materials derived from the anchoring of a well-defined molecular catalyst on a suitable support [17,18]. The latter strategy should guarantee a very high percentage of active sites (potentially 100%), increased robustness of catalyst, mainly due to the inhibition of associative deactivation processes, and possible beneficial cooperation between the anchored catalyst and support. Many hybrid heterogenized catalysts have been reported in the literature [17,19–23] and, among them, those using MOF as support, pioneered by Lin and co-workers [24,25], have been particularly successful [26]. MOFs are a class of porous crystalline compounds constituted by the ordered connection of metal clusters and organic linkers, forming accessible pores and channels potentially useful for a plethora of applications, especially in catalysis and energy production [27–29], as support for metal nanoparticles [30] and for enhanced gas sorption properties [31]. Some of them are rapidly approaching the industrial world [32]. Zr-MOFs are particularly interesting for their chemical and thermal stability and low production cost. The archetype structure is that of UiO-66 which is constituted of hexanuclear clusters with the formula $Zr_6O_4(OH)_4(BDC)_6$ (BDC = 1,4-benzenedicarboxylic acid) in a cubic framework with face-centered cubic (fcu) topology [33]. UiO-66, together with other Zr-MOFs with different topologies, such as NU-1000, was already employed as support for Ir-based catalytic complex, employing both the Zr-cluster site [34,35] and via post-synthetic modification of the ligands for the in situ formation of the complex [36]. UiO-66 possess the exotic feature to be defective when it is crystallized in the presence of a mono carboxylic modulator such as formic, acetic, or benzoic acid, which act as substituent of BDC linker attached to the Zr_6 cluster thus inducing missing linker or missing cluster defects into the structure. These defects can be considered as an opportunity to be employed for imparting targeted functionality to the MOFs by means the so called post-synthetic defect exchange (PSDE) of the monocarboxylic groups with other carboxylic linkers [37,38].

Herein, we report on the synthesis of formic acid (FA)-modulated UiO-66 with a high concentration of defects and its use a support for anchoring, by means of PSDE, an Ir(III) WOC complex based on EDTA (EDTA = *N,N,N′N′*-ethylenediaminotetraacetic acid). The complex, of formula [Ir(HEDTA)Cl]Na (see Figure 1) was already reported in literature as an efficient and durable homogeneous WOC under chemical oxidation with CAN as a sacrificial agent [39]. The molecular structure of the complex clearly shows that it possesses a free carboxylic group which could be employed as an anchoring functionality for its deposition onto a solid surface. Our approach here consists in a PSDE of the FA-UiO-66 MOF with the $[Ir(HEDTA)Cl]^-$ complex dissolved in water. The anchoring occurred through a topotactic exchange of the coordinated FA with the carboxylic group of the IrCl-EDTA complex. The hybrid material (IrEDTA@UiO-66) was characterized by means of surface area and porosity studies, inductively coupled plasma-optical emission spetrometry (ICP-OES) analysis, nuclear magnetic resonance (NMR) spectroscopy, and tested for WO reaction by using CAN as the sacrificial agent. The hybrid exhibited WO activity with TOF and TON values comparable to those of the best performing materials. However, a significant Ir leaching was observed not only during the first catalytic run, as usually observed, suggesting that strong oxidative conditions with Ce^{4+} lead to a rapid decomposition of the hybrid material.

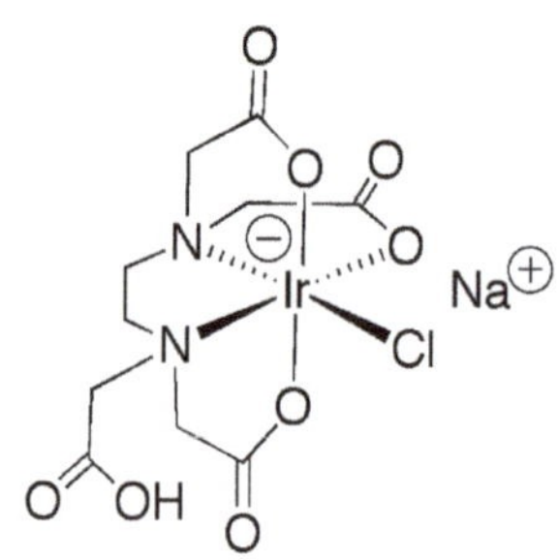

Figure 1. Molecular structure of [Ir(HEDTA)Cl]Na complex.

2. Results and Discussion

2.1. Synthesis and Characterization

Synthesis of FA-Modulated UiO-66

FA-modulated UiO-66 was prepared according to the procedure reported by Taddei et al. [38]. The use of a large amount of formic acid as modulator (100 eq. with respect to Zr) induced the formation of a highly defective phase with respect to the defective free UiO structure, which can be obtained following other synthetic strategies present in literature [40]. FA acts as monocarboxylic modulator with the Zr clusters inducing two types of defects: missing linker defects (Figure 2b) and missing cluster defects (Figure 2c).

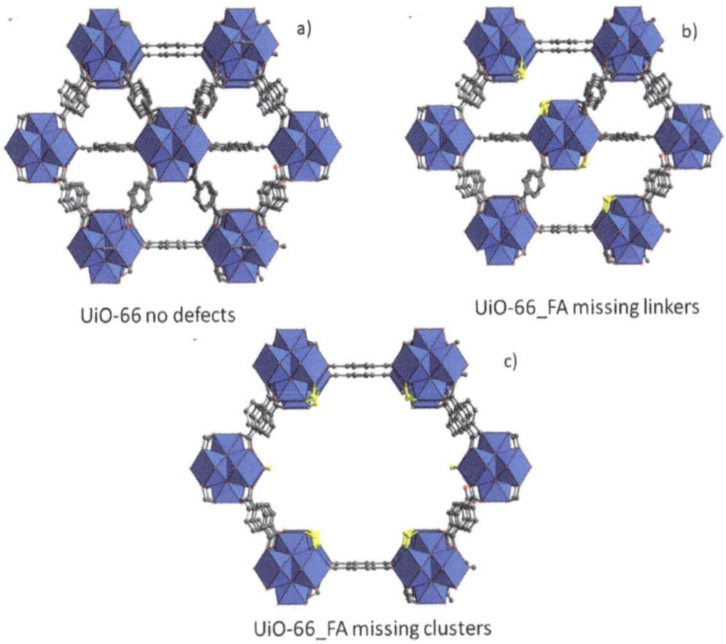

Figure 2. Structure of non-defective (**a**), missing linker defective (**b**), and missing cluster defective (**c**) UIO-66 phase. Formic acid is evidenced in yellow. Zirconium clusters are depicted in blue.

It is known that the materials obtained with FA as modulator most likely possess missing cluster defects [38]. Nitrogen adsorption and desorption analysis at 77 K was performed on FA-UiO-66 compound after activation at 120 °C overnight. The N_2 adsorption/desorption isotherm is reported in Figure 3a and the Brunauer-Emmett-Teller (BET) value is 1450 m^2/g with a total micropore volume of 0.57 cm^3/g. These values, quite higher than the normal surface area and micropore volume of a defect-free UiO-66 (about 1100 m^2/g and 0.4 cm^3/g), suggest the highly defective nature of the obtained material. X-ray powder diffraction (XRPD) pattern of FA_UiO-66 (Figure 3b) shows the peaks at 7.3° and 8.5° of 2θ belonging to the (111) and (200) of the **fcu** UiO-66 phase and a good crystallinity degree. ^{1}H-NMR spectra on the hydrolyzed compound confirm the presence of a considerable amount of FA, as can be seen in Figure 3d. Integration of ^{1}H-NMR signals belonging to FA (8.3 ppm) and BDC (7.8 ppm) gives a FA/BDC ratio equal to 0.63. The obtained solution after the hydrolysis of the sample with NaOH was analyzed with ion chromatography resulting in the following BDC and FA contents in the starting solid: BDC = 2.78 mmol/g and FA = 1.72 mmol/g.

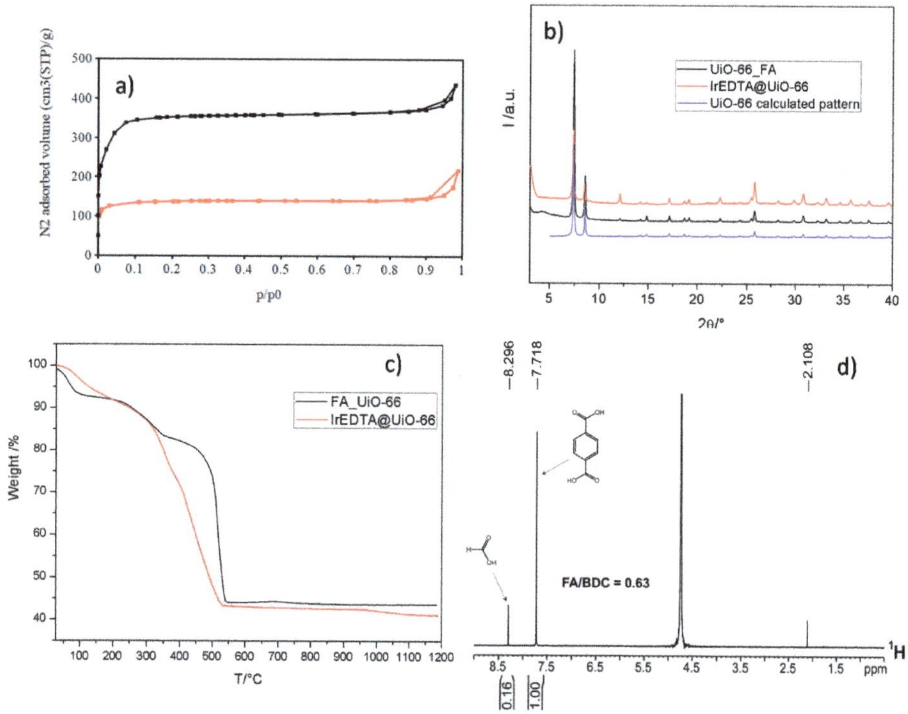

Figure 3. N_2 adsorption and desorption isotherm for FA_UiO-66 (black line) and IrEDTA@UiO-66 (red line) (**a**). XRPD patterns of FA_UiO-66 (black), Ir-EDTA@UiO-66 (red), and calculated pattern for UiO-66 (blue) (**b**). TGA curve for FA_UiO-66 (black) and IrEDTA@UiO-66 (red) (**c**) and ^{1}H-NMR spectrum for hydrolyzed FA_UiO-66 MOF (NaOD/D_2O, 298 K) (**d**).

Given these results, the ratio FA/BDC = 0.63 is in very good agreement with the results of NMR experiments. Since FA is a monocarboxylic acid, the following equation can be used in order to determine the formula of the defective MOF:

$$\frac{FA}{BDC} = \frac{2x}{6-x} = 0.62$$

resulting in $Zr_6O_4(OH)_4(BDC)_{4.58}(FA)_{2.74}$. Thermogravimetric analysis (Figure 3c) shows three different weight losses at 100 °C (7.5%), 330 °C (11%), and 540 °C (38%) due to the loss of water molecules and decomposition of the organic part of the MOF. If the plateau in the 550–1200 °C temperature range is assumed to be $6ZrO_2$ (Mw = 123 g/mol), we can use this value as a reference (100%) for extrapolating the theoretical formula from the analysis. The normalized weight at 100 °C is therefore 213%. The experimental formula weight from TGA analysis at 100 °C is, therefore, 1572 g/mol. Given the mass of the defective, desolvated MOF of formula $Zr_6O_4(OH)_4(BDC)_{4.58}(FA)_{2.74}$ = 1555 g/mol, this is in good agreement with the experimental data from TGA analysis. The PSDE process for anchoring the Ir-EDTA complex onto the cluster surface is shown in Figure S3. After soaking the evacuated MOF into a water solution containing the dissolved complex (0.02 M) and heating at 80 °C for 24 h, the partial exchange of FA with the free carboxylic group of the complex occurred. Three samples with different amounts of exchanged Ir-EDTA were prepared: The compound was exchanged with 0.096, 0.077, and 0.057 mmol of Ir-EDTA, respectively. ICP-OES analysis for the determination of Ir content gave the following results: IrEDTA@UiO-66(1) = 256 µmol/g; IrEDTA@UiO-66(2) = 226 µmol/g; and IrEDTA@UiO-66(3) =170 µmol/g. Figure 3a shows the nitrogen adsorption and desorption analysis at 77 K performed in the same conditions on IrEDTA@UiO-66 containing 256 μmol/g (red curve). After the exchange the calculated BET value is reduced to 547 m^2/g and the total micropore volume is reduced to 0.22 cm^3/g suggesting that the complex is not simply linked to the particle surface but most likely occupies the micropores created by the defects. However, the Ir-EDTA complex is inserted in the micropores and the complex could obstruct a part of the micropore volume of the substrate. TGA analysis of IrEDTA@UiO-66 (Figure 3b, red curve) is similar to that of the pristine MOF although the decomposition of the material starts at lower temperature (around 350 °C) with respect to the unmodified MOF.

The XRPD patterns of the three samples are shown in Figure 4. Anchoring Ir-EDTA onto the cluster surface did not affect the structure of the MOF since the characteristic peaks remained unaltered.

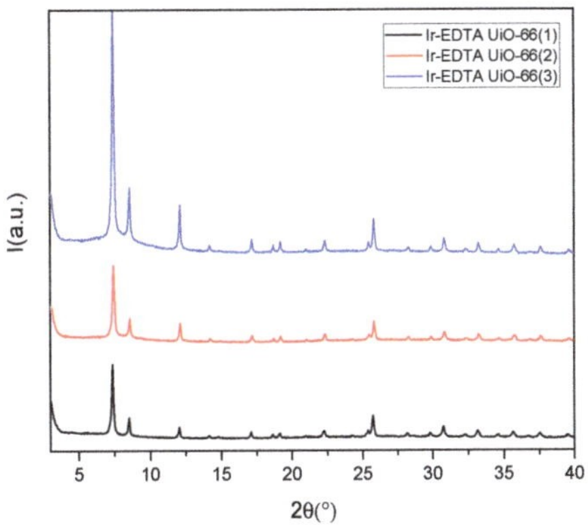

Figure 4. XRPD pattern of IrEDTA@UiO66(1) black, (2) red and (3) blue.

Figure 5 shows the ^{1}H-NMR spectrum of the hydrolyzed IrEDTA@UiO-66 sample. The peak at 8.3 ppm attributed to FA exhibits a reduced intensity and the integration with that of BDC gave as result FA/BDC = 0.10. This value is about six times lower than the unmodified defective MOF (FA/BDC = 0.63) meaning that the most part of FA was successfully exchanged by Ir-EDTA complex. Peaks belonging to the Ir-EDTA complex are clearly visible at 3 and 2.2 ppm. With this new ratio, a suggested formula can be $Zr_6O_4(OH)_4(BDC)_{4.58}(FA)_{0.6}(Ir\text{-}EDTA)_{2.2}$.

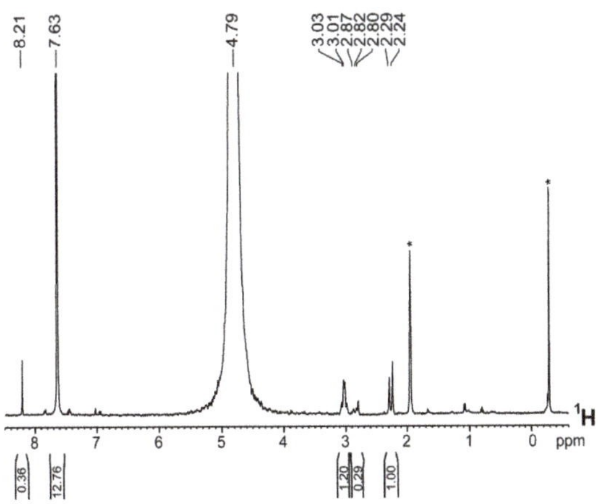

Figure 5. ^{1}H-NMR spectrum of hydrolyzed Ir@UiO-66(1) sample [NaOD/D_2O, 298 K; * denote impurities present in the solvent, likely acetone (ca. 2 ppm) and Silicon Grease (slightly lower than 0 ppm)].

2.2. Water Oxidation Catalytic Activity of IrEDTA@UiO-66

Herein, the catalytic activity of IrEDTA@UiO-66 hybrid materials toward water oxidation to molecular oxygen (Equation (1)) is described. First, a blank experiment by using only the FA_UiO-66 without Ir was performed by adding a 25 mM solution of CAN to 3 mg of MOF (see Figures S1 and S2). No oxygen evolution was observed confirming the inactivity of MOF toward water oxidation. Catalytic tests with Ir containing MOF were carried out by using Ce^{4+} (added as CAN) as a sacrificial oxidant, dispersing the proper amount of catalyst in acidic water (pH 1, 0.1 M HNO_3) at 25 °C.

$$4Ce^{4+} + 2H_2O \rightarrow 4Ce^{3+} + 4H^+ + O_2 \quad (1)$$

The evolved gas, according to Equation (1), was quantified by differential manometry (See Materials and Methods). In a first series of experiments, a consecutive triple addition (100, 150, and 500 μL) of a 1.25 M solution of CAN to 4.9 mL of a 51.5 μM IrEDTA@UiO-66 suspension was executed (Table 1, entries 1–3; Figure 6). IrEDTA@UiO-66 was found to be a competent catalyst for water oxidation and exhibited a TOF of ca. 5 min^{-1} and TON values included between 62 and 308 with yields = 30%–50%. A second series of measurements was performed with the aim of evaluating possible leaching of the molecular catalyst from the MOF support. Particularly, a catalytic run was executed by using 73.12 μM IrEDTA@UiO-66 and 75 mM CAN (Table 1, entry 4). At the end of O_2 evolution IrEDTA@UiO-66 was recovered by filtration and the supernatant solution tested by the addition of another aliquot of 75 mM CAN (Table 1, entry 5). Moreover, the recovered solid was tested under the same conditions (Table 1, entry 6). At the end of the reaction the solid catalyst was again recovered by filtration and the second supernatant tested (Table 1, entry 7). The measured TOF (4 min^{-1}) and TON (108, yield = 42%) values of the starting IrEDTA@UiO-66 are nicely consistent with those observed in the first series of experiments. Furthermore, the recovered solid IrEDTA@UiO-66 exhibits similar TOF (6 min^{-1}) and TON (180, yield = 67%) values. Nevertheless, the two supernatants are active, with even higher TOF (10 and 13 min^{-1}) but comparable TON (363, yield = 44% and 1013, yield = 46%) values, evidencing some leaching of iridium in solution. ICP-OES measurements indicate that 30.98% and 30.75% of iridium leached out from IrEDTA@UiO-66 after the first and second catalytic run, respectively. In order to check the stability of the MOF before catalysis we evaluated the Ir leaching by dispersing IrEDTA@UiO-66 in a 0.1 M HNO_3 solution for 2 h, without the addition of CAN. The measured Ir leaching was about 35%, which is similar to that observed in the first catalytic run. It means that the grafted complex is scarcely stable upon acidic conditions. The catalytic activity of IrEDTA@UiO-66 compares well with those of the molecular precursor [34] and hybrid material IrEDTA@TiO_2 [20], tested under similar conditions, in terms of TOF (Table 1, entries 8–10 and 12). The TON values are clearly lower than those observed for the molecular precursor, which provide 100% yield, and somewhat smaller also than those of IrEDTA@TiO_2 (Table 1, entries 8–10 and 12). Nevertheless, the main criticality of IrEDTA@UiO-66 seems to be the leaching of iridium, occurring also after the second catalytic run, contrary to what observed for IrEDTA@TiO_2 (Table 1, entries 11 and 13) and other heterogenized iridium catalysts reported before [17,20]. Several explanations might be provided for such a phenomenon. It can be hypothesized some Ce^{4+} might undergo an exchange with the Zr^{4+} ions of MOF, becoming not available anymore for driving the oxidative splitting of water. Alternatively, it might be hypothesized that the oxidative potential of iridium inIrEDTA@UiO-66 is slightly higher than in the molecular precursor and hybrid material IrEDTA@TiO_2, thus asking for a higher Ce^{4+}/Ce^{3+} ratio in order to reach the appropriate "Nernstian" potential for WO [41,42]. Both the explanations are consistent with the observation that the addition of a second aliquot of CAN restores the catalytic activity.

Table 1. Summary of the water oxidation (WO) catalytic data for Equation (1). "Sur" indicates supernatant.

Entry	Catalytic Run	[Ir]	[CAN]	d[O_2]/dt	TOF	TON	Yield
		μM	mM	mM/min	min^{-1}		%
	IrEDTA@UiO-66						
1	Run I	50	25	0.23	5	67	54
2	Run II	49	38	0.19	4	32	33
3	Run III	45	126	0.29	6	382	55
4	Run I	73	75	0.31	4	108	42
5	Sur I	23	75	0.22	10	363	44
6	Run II	70	75	0.41	6	180	67
7	Sur II	9	75	0.11	13	1013	46
	IrEDTA [39]						
8	Run I	5	80		7	4000	100
9	Run II	5	20		7	1000	100
	IrEDTA@TiO$_2$ [20]						
10	Run I	35	9		4	46	70
11	Sur I	10	9		7	141	59
12	Run II	25	9		4	78	83
13	Sur II	0	10		–	–	–
14	**IrCp*@Zr-bpy-dc** [25]						
15	Run I	10	3		0.52		17
16	Run II	10	3		0.54		17
17	Run III	10	3		0.53		17
18	**IrCp*@Zr-ppy-dc** [25]						
19	Run I	10	3		6.3		27
20	Run II	10	3		9.4		80

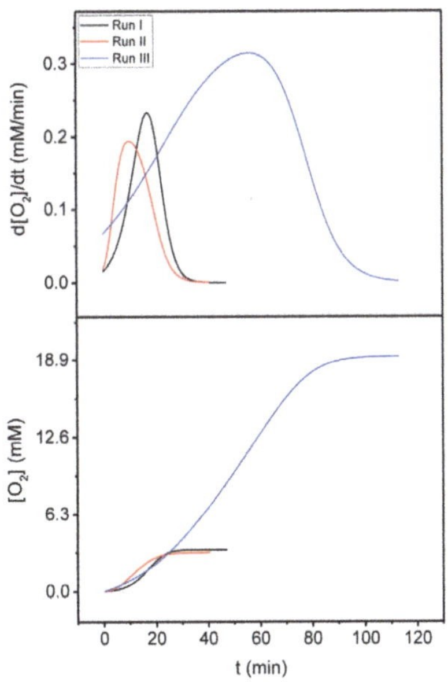

Figure 6. [O_2] (**bottom**) and d[O_2]/dt (**up**) versus time trends for a WO triple cerium ammonium nitrate (CAN) addition experiment (Table 1, entries 1–3).

A catalytic run with a large amount of IrEDTA@UiO-66 (50 mg, 2.61 mM; CAN = 75 mM) was performed in order to recover and analyze IrEDTA@UiO-66 post-catalysis. The ^{1}H-NMR spectrum of the recovered solid digested in NaOD is significantly different than that before catalysis (Figure 7).

In particular, the typical resonances of the $–CH_2$ protons of EDTA in the 2.0–3.2 ppm range are not visible anymore in the post-catalysis sample, suggesting a complete degradation of the ligand framework [43]. XRPD pattern of the MOF after three catalytic runs (Figure S4) shows no crystallinity loss. The FA/BDC ratio (Figure S5) post catalysis is 0.20 suggesting that the framework remained most likely unaltered and the degradation involved a small fraction of BDC together with the Ir-EDTA complex. However, because the recovered solid is still active in WO, it might be hypothesized that after EDTA degradation some iridium remains attached at the MOF structure, possibly through the formation of Zr–O–Ir oxo bridges, as observed in heterogenized WOCs prepared by anchoring an Ir-Kläui molecular precursor onto $BiVO_4$ nanopyramids [17].

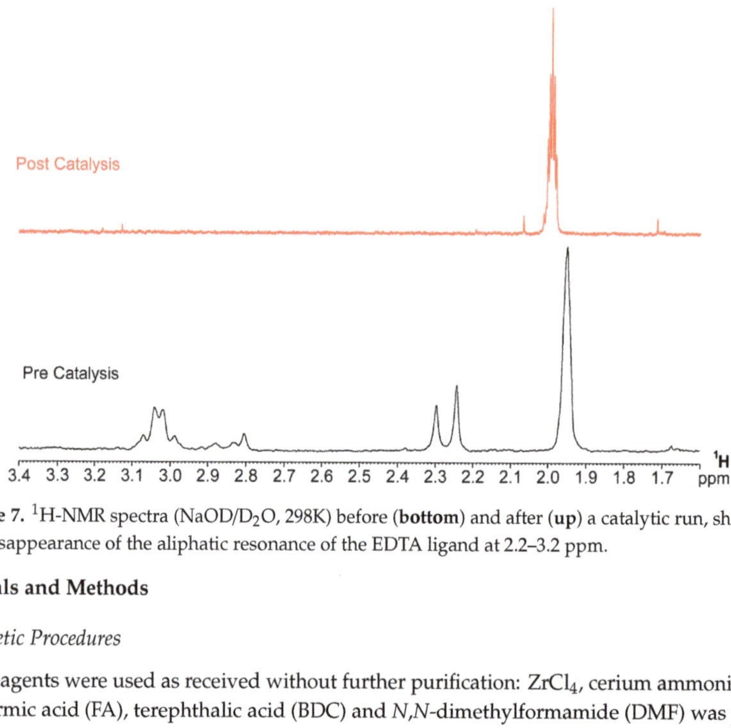

Figure 7. ^{1}H-NMR spectra (NaOD/D_2O, 298K) before (**bottom**) and after (**up**) a catalytic run, showing the disappearance of the aliphatic resonance of the EDTA ligand at 2.2–3.2 ppm.

3. Materials and Methods

3.1. Synthetic Procedures

All reagents were used as received without further purification: $ZrCl_4$, cerium ammonium nitrate (CAN), formic acid (FA), terephthalic acid (BDC) and *N*,*N*-dimethylformamide (DMF) was purchased from Sigma Aldrich (St. Louis, MO, USA). [Ir(HEDTA)Cl]Na was prepared according to Reference [34].

3.1.1. Synthesis of FA-UiO-66

$ZrCl_4$ (0.60 g, 2.5 mmol) was dissolved in DMF (40 mL). Then, water (0.135 mL, 7.5 mmol), FA (9.4 mL, 250 mmol), and BDC (0.435 g, 2.5 mmol) were added to the solution. The mixture was sonicated until complete dissolution and divided in four vials (10 mL each) and heated in an oven at 120 °C for 16 h. After the reaction, the solid was recovered for centrifugation and washed with DMF (one time after 2 h soaking), water (2 h soaking), and acetone (one time after 10 min soaking). At the end, the solid was dried in an oven at 80 °C for 2 h.

3.1.2. Synthesis of IrEDTA@UiO-66 via PSDE

FA_UiO66 (60 mg) was suspended in 5 mL of a 0.02 M water solution of a [Ir(HEDTA)Cl]Na (0.02M) for 24 h at 80 °C. After completion of the reaction, the solid was centrifuged and washed with DMF (one time, two-hour soaking), water (two times, two-hour soaking), and acetone (two times, two-hour soaking). The solid was dried in an oven at 80 °C for two hours. Two other syntheses with different Ir contents were carried out: 30 mg of UiO-66 in 0.01M Ir-EDTA solution (5 mL) and 40 g in 0.015 Ir-EDTA solution (5 mL).

3.2. Analytical and Instrumental Procedures

Powder X-Ray Diffraction (PXRD). PXRD patterns were collected in reflection geometry in the 4–40° 2θ range, with a 40 s per step counting time and with a step size of 0.016° on a PANalytical X'PERT PRO diffractometer (Malvern Panalytical Ltd., Malvern, UK), PW3050 goniometer, (Malvern Panalytical Ltd., Malvern, UK) equipped with an X'Celerator detector (Malvern Panalytical Ltd., Malvern, UK) by using the Cu-Kα radiation. The long fine focus (LFF) ceramic tube operated at 40 kV and 40 mA.

Thermogravimetric analysis (TGA). TGA was performed using a Netzsch STA490C thermoanalyzer (NETZSCH Group, Selb, Germany) under a 20 mL min^{-1} air flux with a heating rate of 10 °C min^{-1}.

Nitrogen adsorption and desorption isotherms. N_2 adsorption/desorption isotherms were performed using a Micromeritics ASAP 2010 analyzer (Micromeritics, Norcross, GA, USA). Prior of the analysis, the samples were degassed overnight under vacuum at 120 °C. BET analysis and t-plot analysis of the adsorption data were used to calculate specific surface area and micropore volume respectively. The Harksin and Jura equation was used as reference for the statistical thickness calculation.

Ion-Chromatography Analysis. Ion chromatography was carried out using a Dionex 500 (Dionex Corp., Sunnyvale, CA, USA) apparatus with a CD20 suppressed conductivity module. Sample analysis was performed as follow: About 30 mg of sample was dispersed in 40 mL of NaOH 0.0125 M and refluxed for 2 h. After reflux, the solution was diluted to 100 mL by water. The resulting solution was analyzed by ion chromatography using a Dionex AS11 column and eluted with a flux of 1.5 mL/min with NaOH 6 mM in the case of BDC analysis or NaOH 0.1 mM in the case of FA analysis.

ICP-OES Analysis. The ICP-OES analysis was carried out using a Varian 700-ES series (Agilent Technologies, Santa Clara, CA, USA) with a standard (2,5,7, and 10 mg/L respectively) of Iridium solution.

WO catalytic experiments. Catalytic experiments were performed using two homemade jacketed glass reactors coupled to a Testo 521-1 manometer. In a typical catalytic run, IrEDTA@UiO-66 suspended in a 0.1 M HNO_3 solution was loaded into the first reaction vessel (working cell), whereas an equal amount of neat water was loaded into the second one (reference cell). Both reactors were sealed with a rubber septum, connected to the manometer, kept at a constant temperature of 25 °C, and placed under stirring for 20 min. Acquisition was started. When a steady baseline was achieved, an equal volume of a solution of CAN and neat water were injected into the working cell and reference cell, respectively, to reach a final volume of 5 mL in each reactor. The concentration of the stock solution of CAN was adjusted, depending on the final concentration desired, in order to have a maximum injection volume of 500 μL. The total gas evolved was estimated by measuring the differential pressure between the working and reference cell.

Fitting methodology and kinetic data analyses. All trends of $[O_2]$ evolution versus time were fitted by a composite mathematical function developed by Peters and Baskin (PB) for distinguishing sigmoidal and bilinear growth profiles of plant roots [38]. The derivative of the PB fits provided reaction rate ($v = d[O_2]/dt$) trends as function of time. Reaction rate over catalyst concentration led to TOF (= v/[Ir]), which was plotted versus the factor conversion X ($= 4[O_2]/[CAN]_0$) [9].

4. Conclusions

In this paper a catalytic active Ir complex based on EDTA was successfully anchored onto a defective Zr-MOF with UiO-66 structure. The post-synthetic modification of defective MOF for designing a new heterogenous catalyst was here validated for the first time demonstrating that substitution of small formate anions linked to zirconium clusters with a larger carboxylate-bearing complex is possible. The material was employed for water oxidation reaction using Ce^{4+} as the sacrificial agent. The catalyst showed a good catalytic activity, which is comparable to that observed for already reported iridium-supported Zr-MOF [25] and slightly lower than Ir-EDTA@TiO2 heterogenized catalysts [20]. However, Ir leaching occurs not only during the first catalytic run, as usually observed, but also for the successive ones. Moreover, leaching of Ir was also observed simply dispersing the solid in the nitric acid solution, without CAN addition. This fact suggests that the material is not stable under the acidic and strong oxidative conditions due to the high redox potential of Ce^{4+}. Furthermore, the WO reaction yield is somewhat lower than that observed for other heterogenized iridium WOCs, indicating a possible exchange of the zirconium atom of MOF with cerium of CAN or a higher "Nernstian" potential. Despite those drawbacks, the results reported in this paper suggest that anchoring a molecular WOC onto a defective MOF is a viable strategy to assemble a hybrid material to be integrated into a device for the generation of renewable fuels. Future developments of this work will be devoted to the stability improvement of the system by performing photo- or electro-catalysis which avoid the use of Ce^{4+} and strong acidic conditions.

Supplementary Materials: The following are available online at http://www.mdpi.com/2304-6740/7/10/123/s1, Figure S1: Manometric oxygen evolution of IrEDTA@UiO-66 and UiO-66. Figure S2: Differential manometric oxygen evolution of IrEDTA@UiO-66 and UiO-66. Figure S3: PSDE of FA with IrEDTA complex onto the structure of FA_UiO-66. Figure S4: XRPD patterns of IrEDTA@UiO-66(3) before and after three catalytic runs. Figure S5: ^{1}H-NMR spectrum of IrEDTA@UiO-66 after 3 catalytic runs.

Author Contributions: Conceptualization, F.C. and A.M.; methodology, F.C., F.M., A.M.; formal analysis, G.G., R.B. and F.M.; data curation, G.G., F.M. and F.C.; writing—original draft preparation, F.C. and A.M.; writing—review and editing, F.C. and A.M.

Funding: This research received no external funding.

Conflicts of Interest: The authors declare no conflict of interest.

References

1. Alstrum-Acevedo, J.H.; Brennaman, M.K.; Meyer, T.J. Chemical Approaches to Artificial Photosynthesis. 2. *Inorg. Chem.* **2005**, *44*, 6802–6827. [CrossRef] [PubMed]
2. Lewis, N.S.; Nocera, D.G. Powering the planet: Chemical challenges in solar energy utilization. *Proc. Natl. Acad. Sci. USA* **2006**, *103*, 15729–15735. [CrossRef] [PubMed]
3. Balzani, V.; Credi, A.; Venturi, M. Photochemical Conversion of Solar Energy. *ChemSusChem* **2008**, *1*, 26–58. [CrossRef] [PubMed]
4. Llobet, A. *Molecular Water Oxidation Catalysis: A Key Topic for New Sustainable Energy Conversion Schemes*; Wiley-Interscience: New York, NY, USA, 2014.
5. Macchioni, A. The Middle-Earth between Homogeneous and Heterogeneous Catalysis in Water Oxidation with Iridium. *Eur. J. Inorg. Chem.* **2019**, *2019*, 7–17. [CrossRef]
6. Corbucci, I.; Macchioni, A.; Albrecht, M. Iridium Complexes in Water Oxidation Catalysis. In *Iridium(III) in Optoelectronic and Photonics Applications*; John Wiley & Sons Ltd.: Hoboken, NJ, USA, 2017; pp. 617–654.
7. Thomsen, J.M.; Huang, D.L.; Crabtree, R.H.; Brudvig, G.W. Iridium-based complexes for water oxidation. *Dalt. Trans.* **2015**, *44*, 12452–12472. [CrossRef] [PubMed]
8. Menendez Rodriguez, G.; Bucci, A.; Hutchinson, R.; Bellachioma, G.; Zuccaccia, C.; Giovagnoli, S.; Idriss, H.; Macchioni, A. Extremely Active, Tunable, and pH-Responsive Iridium Water Oxidation Catalysts. *ACS Energy Lett.* **2017**, *2*, 105–110. [CrossRef]
9. Macchioni, A.; Menendez Rodriguez, G.; Gatto, G.; Zuccaccia, C. Benchmarking Water Oxidation Catalysts Based on Iridium Complexes: Clues and Doubts on the Nature of Active Species. *ChemSusChem* **2017**, 4503–4509.
10. Michaelos, T.K.; Shopov, D.Y.; Sinha, S.B.; Sharninghausen, L.S.; Fisher, K.J.; Lant, H.M.C.; Crabtree, R.H.; Brudvig, G.W. A Pyridine Alkoxide Chelate Ligand That Promotes Both Unusually High Oxidation States and Water-Oxidation Catalysis. *Accounts Chem. Res.* **2017**, *50*, 952–959. [CrossRef] [PubMed]
11. Yang, K.R.; Matula, A.J.; Kwon, G.; Hong, J.; Sheehan, S.W.; Thomsen, J.M.; Brudvig, G.W.; Crabtree, R.H.; Tiede, D.M.; Chen, L.X.; et al. Solution Structures of Highly Active Molecular Ir Water-Oxidation Catalysts from Density Functional Theory Combined with High-Energy X-ray Scattering and EXAFS Spectroscopy. *J. Am. Chem. Soc.* **2016**, *138*, 5511–5514. [CrossRef]

12. Woods, J.A.; Lalrempuia, R.; Petronilho, A.; McDaniel, N.D.; Müller-Bunz, H.; Albrecht, M.; Bernhard, S. Carbene iridium complexes for efficient water oxidation: Scope and mechanistic insights. *Energy Environ. Sci.* **2014**, *7*, 2316–2328. [CrossRef]
13. Venturini, A.; Barbieri, A.; Reek, J.N.H.; Hetterscheid, D.G.H. Catalytic Water Splitting with an Iridium Carbene Complex: A Theoretical Study. *Chem. A Eur. J.* **2014**, *20*, 5358–5368. [CrossRef] [PubMed]
14. McDaniel, N.D.; Coughlin, F.J.; Tinker, L.L.; Bernhard, S. Cyclometalated Iridium(III) Aquo Complexes: Efficient and Tunable Catalysts for the Homogeneous Oxidation of Water. *J. Am. Chem. Soc.* **2008**, *130*, 210–217. [CrossRef] [PubMed]
15. Li, M.; Takada, K.; Goldsmith, J.I.; Bernhard, S. Iridium(III) Bis-Pyridine-2-Sulfonamide Complexes as Efficient and Durable Catalysts for Homogeneous Water Oxidation. *Inorg. Chem.* **2016**, *55*, 518–526. [CrossRef] [PubMed]
16. Fagiolari, L.; Scafuri, A.; Costantino, F.; Vivani, R.; Nocchetti, M.; Macchioni, A. A Ternary Zn–Al–Ir Hydrotalcite-Like Compound Exhibiting High Efficiency and Recyclability as a Water Oxidation Catalyst. *ChemPlusChem* **2016**, *81*, 1060–1063. [CrossRef]
17. Wan, X.; Wang, L.; Dong, C.L.; Rodriguez, G.M.; Huang, Y.-C.; Macchioni, A.; Shen, S. Activating Kläui-Type Organometallic Precursors at Metal Oxide Surfaces for Enhanced Solar Water Oxidation. *ACS Energy Lett.* **2018**, *3*, 1613–1619. [CrossRef]
18. Sheehan, S.W.; Thomsen, J.M.; Hintermair, U.; Crabtree, R.H.; Brudvig, G.W.; Schmuttenmaer, C.A. A molecular catalyst for water oxidation that binds to metal oxide surfaces. *Nat. Commun.* **2015**, *6*. [CrossRef] [PubMed]
19. Chen, Z.; Concepcion, J.J.; Hu, X.; Yang, W.; Hoertz, P.G.; Meyer, T.J. Concerted O atom–proton transfer in the O–O bond forming step in water oxidation. *Proc. Natl. Acad. Sci. USA* **2010**, *107*, 7225–7229. [CrossRef] [PubMed]
20. Savini, A.; Bucci, A.; Nocchetti, M.; Vivani, R.; Idriss, H.; Macchioni, A. Activity and Recyclability of an Iridium–EDTA Water Oxidation Catalyst Immobilized onto Rutile TiO_2. *ACS Catal.* **2015**, *5*, 264–271. [CrossRef]
21. Pastori, G.; Wahab, K.; Bucci, A.; Bellachioma, G.; Zuccaccia, C.; Llorca, J.; Idriss, H.; Macchioni, A. Heterogenized Water Oxidation Catalysts Prepared by Immobilizing Kläui-Type Organometallic Precursors. *Chem. A Eur. J.* **2016**, *22*, 13459–13463. [CrossRef]
22. Materna, K.L.; Rudshteyn, B.; Brennan, B.J.; Kane, M.H.; Bloomfield, A.J.; Huang, D.L.; Shopov, D.Y.; Batista, V.S.; Crabtree, R.H.; Brudvig, G.W. Heterogenized Iridium Water-Oxidation Catalyst from a Silatrane Precursor. *ACS Catal.* **2016**, *6*, 5371–5377. [CrossRef]
23. Materna, K.L.; Crabtree, R.H.; Brudvig, G.W. Anchoring groups for photocatalytic water oxidation on metal oxide surfaces. *Chem. Soc. Rev.* **2017**, *46*, 6099–6110. [CrossRef] [PubMed]
24. Wang, C.; Xie, Z.; Dekrafft, K.E.; Lin, W. Doping Metal–Organic Frameworks for Water Oxidation, Carbon Dioxide Reduction, and Organic Photocatalysis. *J. Am. Chem. Soc.* **2011**, *133*, 13445–13454. [CrossRef]
25. Wang, C.; Wang, J.-L.; Lin, W. Elucidating Molecular Iridium Water Oxidation Catalysts Using Metal–Organic Frameworks: A Comprehensive Structural, Catalytic, Spectroscopic, and Kinetic Study. *J. Am. Chem. Soc.* **2012**, *134*, 19895–19908. [CrossRef]
26. Shao, Q.; Yang, J.; Huang, X. The Design of Water Oxidation Electrocatalysts from Nanoscale Metal–Organic Frameworks. *Chem. A Eur. J.* **2018**, *24*, 15143–15155. [CrossRef] [PubMed]
27. Adil, K.; Belmabkhout, Y.; Pillai, R.S.; Cadiau, A.; Bhatt, P.M.; Assen, A.H.; Maurin, G.; Eddaoudi, M. Gas/vapour separation using ultra-microporous metal–organic frameworks: Insights into the structure/separation relationship. *Chem. Soc. Rev.* **2017**, *46*, 3402–3430. [CrossRef] [PubMed]
28. Adil, K.; Chen, Z.; Weseliński, Ł.J.; Belmabkhout, Y.; Eddaoudi, M. A supermolecular building layer approach for gas separation and storage applications: The eea and rtl MOF platforms for CO_2 capture and hydrocarbon separation. *J. Mater. Chem. A* **2015**, *3*, 6276–6281.
29. Escorihuela, J.; Narducci, R.; Compañ, V.; Costantino, F. Proton Conductivity of Composite Polyelectrolyte Membranes with Metal–Organic Frameworks for Fuel Cell Applications. *Adv. Mater. Interfaces* **2019**, *6*, 1801146. [CrossRef]
30. Wu, R.; Qian, X.; Zhou, K.; Liu, H.; Yadian, B.; Wei, J.; Zhu, H.; Huang, Y. Highly dispersed Au nanoparticles immobilized on Zr-based metal–organic frameworks as heterostructured catalyst for CO oxidation. *J. Mater. Chem. A* **2013**, *1*, 14294–14299. [CrossRef]

31. Wu, H.; Chua, Y.S.; Krungleviciute, V.; Tyagi, M.; Chen, P.; Yildirim, T.; Zhou, W. Unusual and highly tunable missing-linker defects in zirconium metal–organic framework UiO-66 and their important effects on gas adsorption. *J. Am. Chem. Soc.* **2013**, *135*, 10525–10532. [CrossRef] [PubMed]
32. Vilela, S.M.F.; Tomé, J.P.C.; Silva, P.; Paz, F.A.A. Multifunctional metal–organic frameworks: From academia to industrial applications. *Chem. Soc. Rev.* **2015**, *44*, 6774–6803.
33. Cavka, J.H.; Jakobsen, S.; Olsbye, U.; Guillou, N.; Lamberti, C.; Bordiga, S.; Lillerud, K.P. A New Zirconium Inorganic Building Brick Forming Metal Organic Frameworks with Exceptional Stability. *J. Am. Chem. Soc.* **2008**, *130*, 13850–13851. [CrossRef] [PubMed]
34. Yang, D.; Odoh, S.O.; Wang, T.C.; Farha, O.K.; Hupp, J.T.; Cramer, C.J.; Gagliardi, L.; Gates, B.C. Metal–organic framework nodes as nearly ideal supports for molecular catalysts: NU-1000-and UiO-66-supported iridium complexes. *J. Am. Chem. Soc.* **2015**, *137*, 7391–7396. [CrossRef] [PubMed]
35. Yang, D.; Odoh, S.O.; Borycz, J.; Wang, T.C.; Farha, O.K.; Hupp, J.T.; Cramer, C.J.; Gagliardi, L.; Gates, B.C. Tuning Zr6 Metal–Organic Framework (MOF) Nodes as Catalyst Supports: Site Densities and Electron-Donor Properties Influence Molecular Iridium Complexes as Ethylene Conversion Catalysts. *ACS Catal.* **2016**, *6*, 235–247. [CrossRef]
36. Pintado-Sierra, M.; Rasero-Almansa, A.M.; Corma, A.; Iglesias, M.; Sanchez, F. Bifunctional iridium-(2-aminoterephthalate)–Zr-MOF chemoselective catalyst for the synthesis of secondary amines by one-pot three-step cascade reaction. *J. Catal.* **2013**, *299*, 137–145. [CrossRef]
37. Taddei, M. When defects turn into virtues: The curious case of zirconium-based metal–organic frameworks. *Coord. Chem. Rev.* **2017**, *343*, 1–24. [CrossRef]
38. Taddei, M.; Wakeham, R.J.; Koutsianos, A.; Andreoli, E.; Barron, A.R. Post-Synthetic Ligand Exchange in Zirconium-Based Metal–Organic Frameworks: Beware of The Defects! *Angew. Chemie Int. Ed.* **2018**, *57*, 11706–11710. [CrossRef] [PubMed]
39. Savini, A.; Bellachioma, G.; Bolaño, S.; Rocchigiani, L.; Zuccaccia, C.; Zuccaccia, D.; Macchioni, A. Iridium-EDTA as an Efficient and Readily Available Catalyst for Water Oxidation. *ChemSusChem* **2012**, *5*, 1415–1419. [CrossRef] [PubMed]
40. De Stefano, M.R.; Islamoglu, T.; Garibay, S.J.; Hupp, J.T.; Farha, O.K. Room-temperature synthesis of UiO-66 and thermal modulation of densities of defect sites. *Chem. Mater.* **2017**, *29*, 1357–1361. [CrossRef]
41. Codolà, Z.; Gamba, I.; Acuña-Parés, F.; Casadevall, C.; Clémancey, M.; Latour, J.-M.; Luis, J.M.; Lloret-Fillol, J.; Costas, M. Design of Iron Coordination Complexes as Highly Active Homogenous Water Oxidation Catalysts by Deuteration of Oxidation-Sensitive Sites. *J. Am. Chem. Soc.* **2019**, *141*, 323–333. [CrossRef] [PubMed]
42. Zuccaccia, C.; Bellachioma, G.; Bortolini, O.; Bucci, A.; Savini, A.; Macchioni, A. Transformation of a Cp*–Iridium(III) Precatalyst for Water Oxidation when Exposed to Oxidative Stress. *Chem. A Eur. J.* **2014**, *20*, 3446–3456. [CrossRef] [PubMed]
43. Peters, W.S.; I Baskin, T. Tailor-made composite functions as tools in model choice: The case of sigmoidal vs bi-linear growth profiles. *Plant Methods* **2006**, *2*, 11. [CrossRef] [PubMed]

inorganics MDPI

Article

Photoluminescent Coordination Polymers Based on Group 12 Metals and 1*H*-Indazole-6-Carboxylic Acid

Antonio A. García-Valdivia [1], Estitxu Echenique-Errandonea [2], Gloria B. Ramírez-Rodríguez [1], José M. Delgado-López [1], Belén Fernández [3], Sara Rojas [1], Javier Cepeda [2,*] and Antonio Rodríguez-Diéguez [1,*]

1 Departamento de Química Inorgánica, Facultad de Ciencias, Universidad de Granada, 18071 Granada, Spain; antonioandresgarval@correo.ugr.es (A.A.G.-V.); gloria@ugr.es (G.B.R.-R.); jmdl@ugr.es (J.M.D.-L.); srojas@ugr.es (S.R.)

2 Departamento de Química Aplicada, Facultad de Química, Universidad del País Vasco/Euskal Herriko Unibertsitatea (UPV/EHU), 20018 Donostia, Spain; estitxu.echenique@ehu.eus

3 Institute of Parasitology and Biomedicine "López-Neyra", CSIC, 18600 Granada, Spain; belenfernandez@ipb.csic.es

* Correspondence: javier.cepeda@ehu.es (J.C.); antonio5@ugr.es (A.R.-D.); Tel.: +34-943015409 (J.C.); +34-958248524 (A.R.-D.)

Abstract: Two new coordination polymers (CPs) based on Zn(II) and Cd(II) and 1*H*-indazole-6-carboxylic acid (H_2L) of general formulae $[Zn(L)(H_2O)]_n$ (**1**) and $[Cd_2(HL)_4]_n$ (**2**) have been synthesized and fully characterized by elemental analyses, Fourier transformed infrared spectroscopy and single crystal X-ray diffraction. The results indicate that compound **1** possesses double chains in its structure whereas **2** exhibits a 3D network. The intermolecular interactions, including hydrogen bonds, C–H$\cdots\pi$ and $\pi\cdots\pi$ stacking interactions, stabilize both crystal structures. Photoluminescence (PL) properties have shown that compounds **1** and **2** present similar emission spectra compared to the free-ligand. The emission spectra are also studied from the theoretical point of view by means of time-dependent density-functional theory (TD-DFT) calculations to confirm that ligand-centred π-π^* electronic transitions govern emission of compound **1** and **2**. Finally, the PL properties are also studied in aqueous solution to explore the stability and emission capacity of the compounds.

Keywords: group 12 metals; 1*H*-indazole-6-carboxylic acid; coordination polymer; photoluminescence properties

Citation: García-Valdivia, A.A.; Echenique-Errandonea, E.; Ramírez-Rodríguez, G.B.; Delgado-López, J.M.; Fernández, B.; Rojas, S.; Cepeda, J.; Rodríguez-Diéguez, A. Photoluminescent Coordination Polymers Based on Group 12 Metals and 1*H*-Indazole-6-Carboxylic Acid. *Inorganics* **2021**, *9*, 20. https://doi.org/10.3390/inorganics9030020

Academic Editor: Andrea Rossin

Received: 16 January 2021
Accepted: 17 March 2021
Published: 22 March 2021

1. Introduction

The study of coordination polymers (CPs) and metal-organic frameworks (MOFs) is at the forefront of modern inorganic chemistry due to their broad range of potential applications, spanning from magnetism and luminescence, through catalysis and sensing, to gas separation and storage, and biomedicine [1,2]. Through an adequate selection of their building blocks (metal ions and organic ligands), CPs and MOFs can be designed to enhance a particular property [3–5]. It is well known that nitrogen-containing heterocycles are molecules commonly employed as ligands owing to not only their good coordination ability, but also pharmacological relevance, given that they are important scaffolds widely present in numerous commercially available drugs [6]. The most famous are diazepam, isoniazid, chlorpromazine, metronidazole, barbituric acid, captopril, chloroquinine, azidothymidine and anti-pyrine. As a result of their diverse biological activities, nitrogen heterocyclic compounds have always been attractive targets to develop new active compounds. This is the case for 1*H*-indazole-6-carboxylic acid (H_2L), a common moiety in the pharmaceutical industry [7]. Polysubstituted indazole-containing compounds furnished with different functional groups usually present significant pharmacological activities and serve as structural motifs in drug molecules (i.e., niraparib-anticancer drug, pazopanib-approved by the FDA for renal cell carcinoma, bendazac and benzydamine-antiinflamatory drugs) [8]. From the structural point of view, indazole is an aromatic heterocyclic molecule with a benzene ring fused to a pyrazole ring [9]. It shows three tautomeric forms (Scheme 1) being tautomer **A** favoured over **B** and **C** due to its higher degree of aromaticity [10].

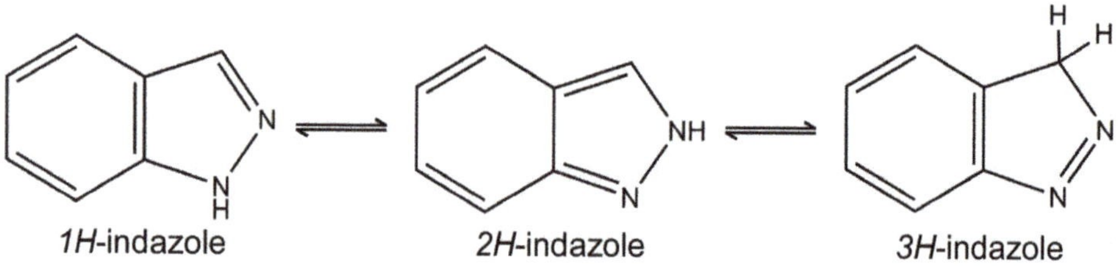

Scheme 1. Indazole tautomerism (**A**, **B**, and **C** from left to right).

H_2L is presented here as an ideal candidate to form CPs or MOFs as it possesses multiple coordination possibilities, not only derived from its carboxylate group, but also from its pyrazole function. Thus, a great variety of coordination modes are possible according to similar ligands containing carboxylate and pyrazole chemical functions in crystallized complexes (Scheme 2). Until now, only one complex based on this ligand has been reported so far [11]. In that work, Kruger et al. described in detail four substituted indazole derivatives containing pyridine or carboxylic functionalities upon coordination with Cu(II) ions in solution and solid state. In the complex, 1*H*-indazole-6-carboxylate acts as a bridging ligand showing a tridentate coordination mode: the carboxylate group coordinates to two Cu(II) atoms in a *syn,syn* mode to establish a dimeric paddle-wheel shaped entity, whereas the non-protonated nitrogen atom of the pyrazole ring links to a third Cu(II) atom in a monodentate way (see the highlighted modes in Scheme 2). Aside from this work mainly focused on the description of a new compound, it should be pointed out that some Co(II)-based complexes with indazole derivatives have shown a capacity to bind to DNA [12].

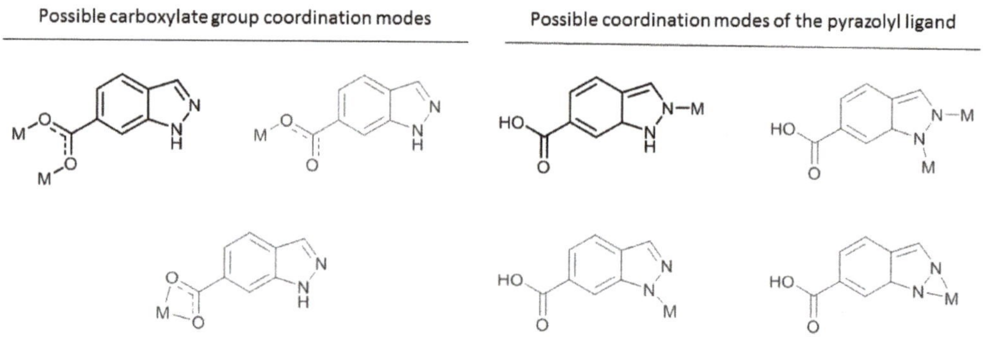

Scheme 2. Possible coordination modes of 1*H*-indazole-6-caboxylate ligand. Note that only those two modes highlighted in black have been described in bibliography whereas the rest correspond to potential binding modes.

On the other hand, H_2L may also present interesting photoluminescence (PL) properties due to its aromatic nature and the presence of carboxylic groups, with potentially strong light absorption [13]. When these indazole-carboxylate ligands are coordinated to metal centres in the crystal structure of a CP, PL tends to be enhanced by means of the well-known crystal-induced luminescence effect [14]. Among others, metal ions from group 12 are particularly appropriate for their use in PL as they present a closed-shell electronic configuration in which d-d transitions cannot occur [15,16]. In fact, many CPs and MOFs formed by these metal ions have been reported during the last decade [17,18], some of which present not only strong and bright fluorescent emissions, but also long-lived phosphorescence that may be traced by the naked eye [19–21]. Moreover, the presence of these ions may also promote ligand-to-metal charge transfer (LMCT) as metal ions possess empty orbitals that can be populated in the excited state, and therefore the PL emission may be modulated with regard to the ligand-centred (LC) emissions [22,23]. Irrespective of the luminescence mechanism occurring in these systems, the interest for group 12-based compounds has increased given their potential application as not only lighting devices, but also as luminescence-based molecular detectors [24], thermometers [25] and anti-counterfeiting inks, among others [26]. Particularly for indazole derivatives playing as ligands, many Zn-/Cd-indazole complexes have already proved efficient luminescent CPs under UV irradiation [27].

Considering all the above, in this work we present the synthesis, structural characterisation and PL properties of two new coordination polymers based on group 12 metals and 1*H*-indazole-6-carboxylic acid of general formula $[Zn(L)(H_2O)]_n$ (**1**) and $[Cd_2(HL)_4]_n$ (**2**). Their emission characteristics have been studied both from the theoretical and experimental points of view, involving the measurements in the solid state as well as in aqueous medium.

2. Results and Discussion

The reaction of 1*H*-indazole-6-carboxylic acid gave rise to two compounds based on group 12 metals which exhibit a different structural dimensionality. In particular, the solvothermal reaction of the 1*H*-indazole-6-carboxylic acid ligand with zinc acetate salt ($Zn(CH_3COO)_2$) using a 1:2 molar ratio in a *N,N*-dimethylformamide/water (DMF/H_2O) mixture afforded a 1D CP, namely **1** (see Experimental Section for further details). Similarly, the use of cadmium acetate salt ($Cd(CH_3COO)_2$) salt in the synthesis, successfully led to a 3D MOF, namely **2**. This fact can be explained by the larger ion size of Cd(II), which may admit higher coordination numbers, involving the participation of additional ligands and increasing the metal-to-ligand connectivity.

2.1. Description of the Structures

2.1.1. Structural Description of $[Zn(L)(H_2O)]_n$ (1)

Compound **1** crystallizes in the $P2_1/n$ space group and consists of a double chain structure in which Zn(II) ions are bridged by nitrogen atoms of L^{2-} in a bidentate way, giving rise to a stable and in plane Zn_2N_4 dimeric core as a six membered ring (Figure 1).

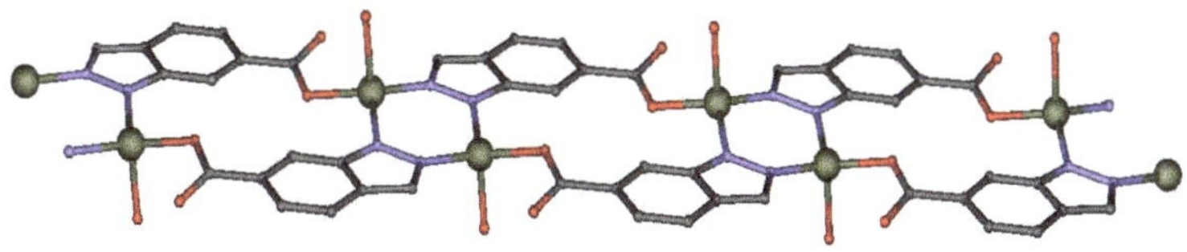

Figure 1. Representation of the 1D polymeric chain in which Zn_2N_4 planar six membered ring is observed (zinc, nitrogen, oxygen, and carbon are represented in green, blue, red, and grey, respectively; hydrogen atoms are omitted for clarity).

The Zn(II) ion is also coordinated to a carboxylate moiety of the indazole derivative ligand in a monodentate way, which extends the dimeric entity into infinite 1D chains running along the crystallographic [100] direction. The coordination sphere of Zn is completed by the coordination of a water molecule (see Table S1 in the ESI for further information about bond lengths and angles). The ZnN_2O_2 coordination sphere can be described as a tetrahedron, although Zn ions show a geometry close to an axially vacant trigonal bipyramid according to continuous-shape-measures (CShMs) using SHAPE software (Tables S2 and S3, in the ESI) [28].

The packing of the double chains is ruled by intermolecular interactions, among which hydrogen bonding interactions established between coordination water molecules and carboxylate oxygen atoms are to be highlighted (Figure 2). In particular, the coordinated water molecule is involved in hydrogen bonding interactions in which non-coordinated carboxylate oxygen atoms belonging to adjacent chains act as receptors. Additionally, the angle formed among neighbouring chains allows for the formation of C–H···π interactions between aromatic rings, reinforcing the stability of the supramolecular crystal building (see Figure S5 in the ESI).

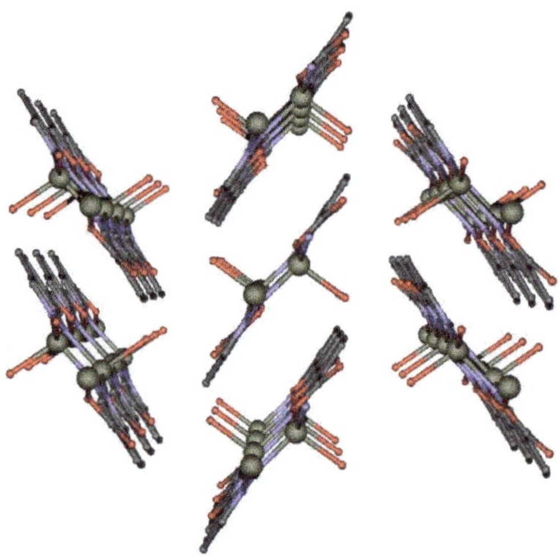

Figure 2. Perspective view of the chains of $[Zn(L)(H_2O)]_n$ packed in the framework (hydrogen atoms have been omitted for clarity).

2.1.2. Structural Description of $[Cd_2(HL)_4]_n$ (**2**)

Compound **2** crystallizes in the triclinic *P*-1 space group. The asymmetric unit contains two non-equivalent Cd(II) atoms and four ligand molecules. Each Cd(II) ion is connected to two monodentate indazole nitrogen atoms and four oxygen atoms of the carboxylate group of the ligand. Cd1 and Cd2 ions are doubly linked by ancillary *syn-anti* carboxylate moieties of 1-*H*-indazole-6-carboxylate ligands (namely A, C and D). However, the carboxylate group of B ligand presents a different coordination pattern, in which O1B connects in a monodentate way to both Cd1 and Cd2 atoms giving rise to alternating five and six membered rings (Figure 3, see also the view along *b* axis in Figure S6 in the ESI), whereas O2B atom remains unconnected to any metal centre.

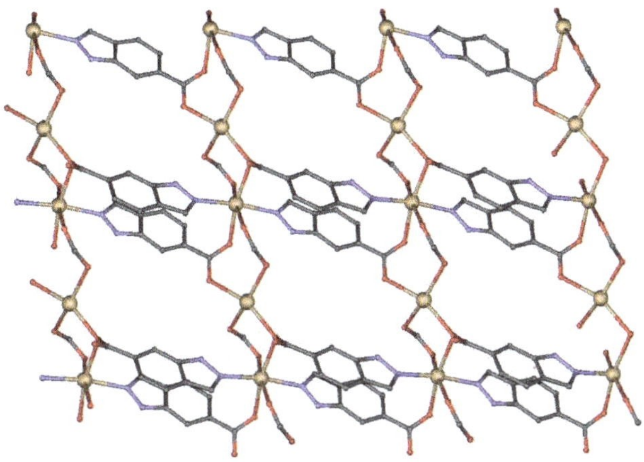

Figure 3. View of the coordination of Cd(II) ions to HL in compound **2** (cadmium, nitrogen, oxygen, and carbon are represented in dark-yellow, blue, red, and grey, respectively; hydrogen atoms have been omitted for clarity).

CShMs indicate that different ligand coordination modes affect the connectivity of the metal centres, which leads to the formation of a distinct crystal structure. When comparing **1** and **2** compounds, the coordination spheres of Cd1 and Cd2 are described as octahedra according to SHAPE measurements (Tables S2 and S3 in the SI). M···N2 distances are slightly shorter than in compound **1** (in the 1.969 and 2.293–2.316 Å ranges, respectively), similarly to the M···$O1_{carboxylate}$ bond distances (between 1.935 and 2.320 Å, see Table S1 in the ESI). As a result, a 3D framework is obtained in the case of **2** by the further linkage of the carboxylate groups to Cd(II) atoms along a metal-carboxylate rod (Figure 4). Considering the connectivity of the metal ions and HL ligands, this framework may be described as a **5,6T24** topological network with the $(3^2.4^2.5^2.6^3.7)_2(3^2.4^4.5^4.6^2.7^3)$ point symbol, as previously observed in the MOF of $[Al_2(OH)_2(H_2O)_2(C_{10}O_8H_2)]$ or MIL-118A [29].

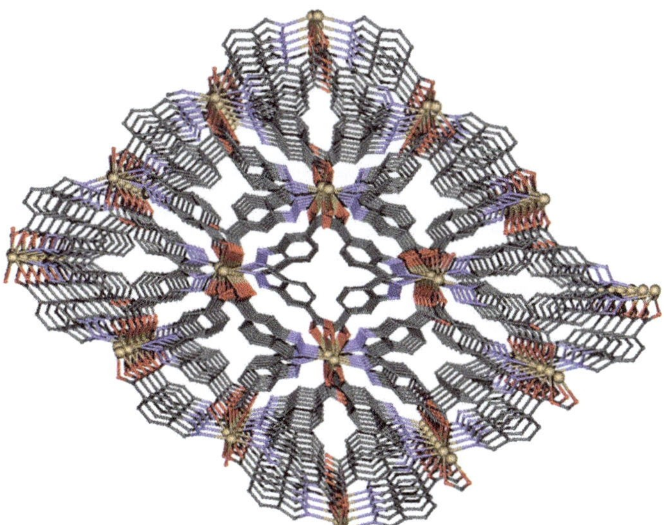

Figure 4. Perspective of 3D structure of compound **2** (hydrogen atoms have been omitted for clarity).

To end up with the structural description, it is worth mentioning that compound 2 presents some remarkable supramolecular interactions that reinforce its packing. Unlike with hydrogen bonding and C–H$\cdots\pi$ interactions governing the crystal structure of **1**, 2 contains $\pi\cdots\pi$ stacking interactions. In particular, the aromatic rings of HL promote strong face-to-face contacts among the whole structure (see Figure S5 in the ESI).

2.2. Fourier Transformed Infrared (FTIR) Spectroscopy

The analysis of the FTIR spectra of **1** and **2** confirms the coordination of zinc(II) and cadmium(II) ions to the *N*-containing carboxylate ligand (Figure S1, see the ESI). FTIR spectra of both compounds confirmed a shift in the wavelengths in comparison to the pure linker, suggesting the formation of interactions between the linker and the metals. The main vibrations of 1*H*-indazole-6-caboxylic acid associated with the $\upsilon_{C=N}$ stretching vibration at 1633 cm^{-1}, and the asymmetric and symmetric vibrations of the carboxylate groups at 1683 and 1423 cm^{-1} are shifted when compared with the spectra of **1** and **2**. The bands found at 1537 and 1589 cm^{-1} for complex **1** and **2**, respectively, are attributed to the $\upsilon_{C=N}$ stretching vibration of the indazole ring [30]. Moreover, the strong absorption peak observed at 1558 and 1402 cm^{-1} for **1**, and 1541 and 1411 cm^{-1} for **2**, respectively, revealed the asymmetric and symmetric vibrations of the carboxylic groups [31]. Finally, the strong broad band in the range of 3317–3086 cm^{-1} was assigned to the O–H stretching vibration of the coordinated water molecule in complex **1** [32].

2.3. Luminescence Properties

As previously mentioned, complexes consisting of metal ions with d^{10} electronic configuration are known to yield strong PL emissions. The completely filled d-orbitals disable ligand field d-d transitions, eliminating fluorescence quenching and allowing the occurrence of PL [14]. Thus, the development of d^{10}-based compounds is interesting for photochemical, electroluminescence and sensing applications [17,33]. The extended aromaticity of the 1*H*-indazole-6-carboxylate ligand coordinated to Zn(II) and Cd(II) atoms suggests the existence of emissive properties of **1** and **2**. The emission of these compounds are found to be similar to ligand emission, which may stem from the ligand-centred π–π* electronic transitions, as shown in Figure 5. Consequently, it can be suggested that the highly conjugated 1*H*-indazole-6-caboxylate ligand is the main part contributing to the emission [34]. An intense broad band at 350–450 nm dominates the emission spectra of all compounds upon 325 nm excitation (in view of the maxima found in the excitation spectra), among which the maxima at 362 and 388 nm, 363 and 381 nm, and 363 and 391 nm can be distinguished for the ligand, and compounds **1** and **2**, respectively; which imbues all compounds with blue emission. The similar emission band of **2** and the free ligand must be attributed to the fact that **2** possesses the protonated form of the ligand (HL$^-$) whereas it is completely deprotonated (L^{2-}) in **1**. In a comparative scale, the ligand spectrum shows two well-defined maxima (not that easily identified for the compounds) and relatively higher intensity (Figure S2, SI). It is worth noticing that the observed luminescence resembles to that shown by other previously reported CPs containing other isomers of indazole-carboxylates [35,36].

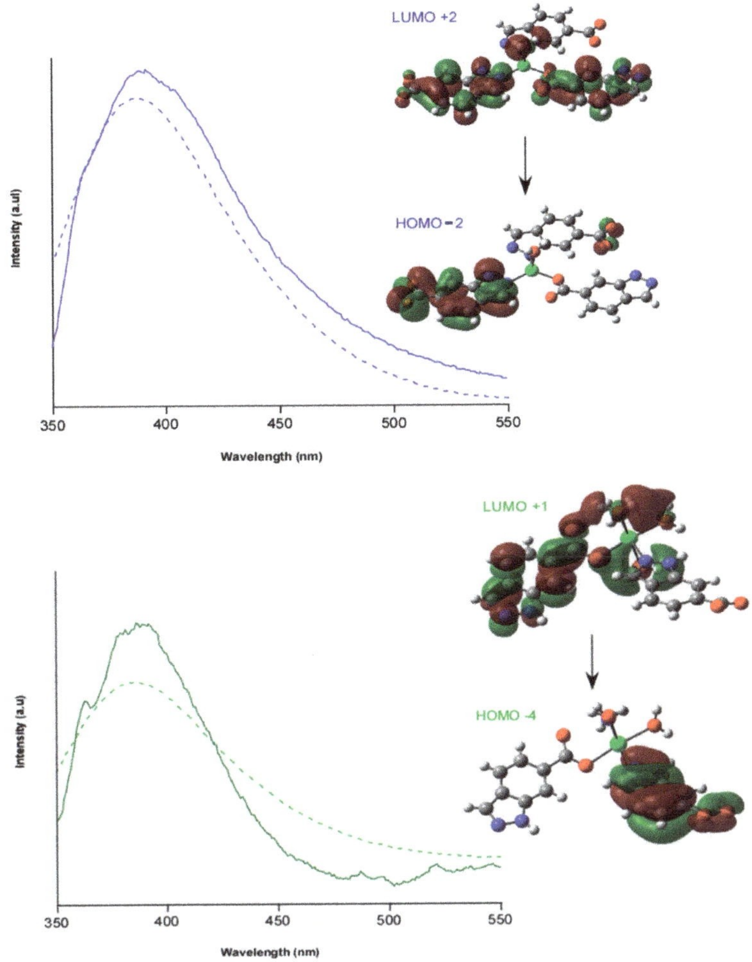

Figure 5. Room temperature Time-dependent density-functional theory (TD-DFT) computed (dashed lines) and experimental (solid line) photoluminescence emission under λ_{ex} = 325 nm of compounds **1** (blue-up) and **2** (green-down). The insets show the most representative molecular orbitals involved in the electronic transitions.

In order to get a deeper insight into the emission mechanism, TD-DFT calculations were performed on suitable models of compounds **1** and **2**. The calculated spectra reproduce fairly well the experimental ones, indicating that the process is driven by singlet transitions occurring between the molecular orbitals shown in Figure 5. In both cases, the electron density in HOMO orbitals, HOMO-2 and HOMO-4 for compound **1** and **2**, respectively, extend over the bonds over the whole ligand molecule (signifying a π orbital) whereas LUMO orbitals, LUMO+2 and LUMO+1 for compound **1** and **2**, respectively, feature a π^* character. Therefore, it can be stated that the transitions involved in the photoluminescence of compound **1** and **2** are mainly of $\pi^* \leftarrow \pi$ nature induced by ligand centred emission.

Inspired by the potential biomedical properties of the ligand on the basis of its similar structure to other indazole derivatives [7,8], we studied the stability and fluorescence performance of these compounds in aqueous media in order to explore their performance as luminescent probes. First, the stability of both compounds was confirmed by recording UV-Vis absorption spectra on aqueous solutions of both compounds immediately after their solution and also after 24 h (see Figure S7 in the ESI). These spectra show the presence of three main absorption bands (sited at ca. 215, 265 and in the 340 nm for **1** and 220, 265 and 305 nm for **2**), corresponding to intraligand and/or ligand-to-metal charge transfers occurring in the complexes. It is worth noticing that these bands are in good agreement with the experimental and TD-DFT computed excitation bands, finding only slight shifts that may be attributed to the different media in which the spectra are recorded (water for UV-Vis and solid state for excitation spectra). Moreover, these solutions were also employed to measure the PL emission spectra of both compounds. As observed in Figure S4, the emission spectra acquired in this medium do not significantly differ from those measured in solid state but for a drop in the intensity of the signal, which is an expected behaviour given that the capacity of this solvent to quench the PL is a largely reported effect [37]. All these results suggest that these new compounds could show potential activity as luminescent probes in some particular biological media (i.e., as biosensor), a fact that *a priori* excludes most of biological tissues owing to their low transparency (high light absorption capacity) to the blue emission (λ_{em} = 350–450 nm) shown by these compounds.

3. Materials and Methods

3.1. Materials and Physical Measurements

All the reagents were purchased commercially and used without any previous purification. Elemental analysis (C, H and N) were carried out at the Centro de Instrumentación Científica (University of Granada) on a Fisons-Carlo Erba analyzer model EA 1108 (Thermo Scientific, Waltham, MA, USA). FTIR spectra (400–4000 cm^{-1}) were recorded on a Nicolet FT-IR 6700 spectrometer (Thermo Scientific, Madrid, Spain) in KBr pellets.

3.2. Synthesis of $[Zn(L)(H_2O)]_n$ (1)

0.010 g (0.006 mmol) of 1*H*-indazole-6-carboxylic acid (H_2L) was dissolved in 0.5 mL of DMF. Then, 0.5 mL of water was added to the ligand solution. In a separate vial, 0.0134 g (0.03 mmol) of $Zn(CH_3COO)_2$ was dissolved in 0.5 mL of water. Similarly, once metal salt was dissolved, 0.5 mL DMF were added to the solution. Metal solution was added dropwise to the ligand solution, and the resulting colourless mixture was placed in a closed glass vessel and heated in an oven at 100 °C for 24 h. X-ray quality crystals of **1** were obtained during heating process under autogenous pressure and washed with water. Yield: 64% based on Zn. Anal Calcd. for $C_8H_4N_2O_3Zn$: C, 39.46; H, 2.48; N, 11.50. Found: C, 39.39; H, 2,41; N, 11.59. In addition to the elemental analyses, the purity of all the samples was checked by FT-IR spectra.

3.3. Synthesis of $[Cd_2(HL)_4]_n$ (2)

The same synthetic procedure was carried out to obtain complex **2**, by replacing $Zn(CH_3COO)_2$ by 0.01651 g (0.03 mmol) of $Cd(CH_3COO)_2$. X-ray quality crystals were obtained and washed with water. Yield: 54% based on Cd. Anal Calcd. for $C_{32}H_{20}Cd_2N_8O_8$: C, 44.21; H, 2.32; N, 12.89. Found: C, 44.16; H, 2.29; N, 12.91. In addition to the elemental analyses, the purity of all the samples was checked by FT-IR spectra.

3.4. Crystallographic Refinement and Structure Solution

Single crystals of suitable dimensions were used for data collection. For compound **1** and **2**, diffraction intensities were recorded on a Bruker X8 APEX II and Bruker D8 Venture with a Photon detector (Bruker, Madrid, Spain) equipped with graphite monochromated MoKα radiation (λ = 0.71073 Å). The data reduction was performed with the APEX2 software [38] and corrected for absorption using SADABS [39]. In all cases, the structures were solved by direct methods and refined by full-matrix least-squares with SHELXL-2018 [40]. The main refinement parameters are listed in Table 1. Details of selected bond lengths and angles are given in Table S2 in the ESI. CCDC reference numbers for the structures are 1,948,382 and 1,948,383 for Cd and Zn coordination polymers, respectively.

Table 1. Crystallographic data and structure refinement details of compounds **1** and **2**.

Compound	1	2
Formula	$C_8H_6N_2O_3Zn$	$C_{32}H_{20}Cd_2N_8O_8$
M_r (g mol^{-1})	243.52	869.36
Crystal system	monoclinic	triclinic
Space group	$P2_1/n$	P–1
Temperature (K)	100 (2)	100 (2)
a (Å)	9.774 (3)	8.7080 (4)
b (Å)	5.7633 (15)	9.0640 (3)
c (Å)	14.592 (4)	19.4510 (7)
α (°)	90	101.089 (1)
β (°)	95.626 (7)	90.961 (2)
λ (°)	90	98.063 (1)
V (Å^3)	818.0 (4)	1490.18 (1)
Z	4	2
ρ (g cm^{-3})	1.977	1.937
μ (mm^{-1})	2.979	1.497
Unique reflections	1035 (817)	7671 (5938)
R_{int}	0.1342	0.0722
GoF [a]	1.082	1.107
R_1 [b]/wR_2 [c] [I>2σ(I)]	0.0777/0.0535	0.0604/0.0352
R_1 [b]/wR_2 [c] [all data]	0.1075/0.0994	0.0823/0.0672
Largest difference in peak and hole (e Å^{-3})	0.635 and −0.0656	1.062 and −1.046

[a] $R_1 = S||F_o| - |F_c||/S|F_o|$. [b] Values in parentheses for reflections with $I > 2s(I)$. [c] $wR_2 = \{S[w(F_o^2 - F_c^2)^2] / S[w(F_o^2)^2]\}^{\frac{1}{2}}$; where $w = 1/[\sigma^2(F_0^2) + (aP)^2 + bP]$ and $P = (max(F_0^2,0) + 2Fc^2)/3$ with a = 0.0.0319 (**1**), 0.0380 (**2**) and b = 6.9969 (**1**).

3.5. Photophysical Measurements

UV-Vis absorption spectra were recorded on UV-2600 UV/vis Shimadzu spectrophotometer using polycrystalline samples of compounds **1** and **2**. PL measurements were carried out on crystalline samples at room temperature using a Varian Cary-Eclipse fluorescence spectrofluorometer equipped with a Xe discharge lamp (peak power equivalent to 75 kW), Czerny–Turner monochromators, and an R-928 photomultiplier tube. For the fluorescence measurements, the photomultiplier detector voltage was fixed at 600 V, and the excitation and emission slits were set at 5 and 2.5 nm, respectively. Phosphorescence spectra were recorded with a total decay time of 20 ms, delay time of 0.2 ms and gate time of 5.0 ms. The photomultiplier detector voltage was set at 800 V, and both excitation and emission slits were open to 10 nm.

4. Conclusions

The reaction between 1*H*-indazole-6-carboxylic acid ligand and Zn(II) or Cd(II) leads the formation of two new coordination polymers with different dimensionalities. Compound **1** possesses a double chain structure, whereas compound **2** exhibits a 3D structure. Emissive properties of both complexes have been studied demonstrating that their photoluminescent emission is driven by the ligand centred $\pi^*\leftarrow\pi$ transition. The similar luminescent properties between compound **2** and the linker may be consequence of the partially protonated HL^- ligand present in **2**. This work is pioneer in studying and comparing the luminescent properties of 1*H*-indazole-6-carboxylic acid (H_2L) and its complexes, which represent a common moiety in pharmaceutical industry. In this regard, novel materials based on this ligand are being developed in our laboratory using lanthanide ions to enhance their luminescent properties.

Supplementary Materials: The following are available online at https://www.mdpi.com/2304-67(…) 0/9/3/20/s1, Figure S1: Infrared spectra of the ligand and compounds **1** and **2**, Figure S2: Emission spectra of the ligand and compounds **1** and **2** under λ_{ex} = 325 nm, Figure S3: Excitation spectra of compounds monitored at the emission maxima: (a) λ_{em} = 381 nm for **1** and (b) λ_{em} = 391 nm for **2**, **Figure S4**: Comparative view of the absorption spectra of compounds (a) **1** and (b) **2** in solid state and aqueous solution, Figure S5: The most representative intermolecular interactions and packing modes for complexes **1** (up) and **2** (down). H bonds, $\pi\cdots\pi$ and C-H$\cdots\pi$ interactions are shown with dashed blue, green and orange lines, respectively, Figure S6: View along a (left), b (middle) and c (right) axis of complex **1** (up) and **2** (down), Figure S7: UV-Vis spectra of compounds (a) **1** and (b) **2** in aqueous solutions acquired at times (0 h and after 24 h), Table S1: Selected bond lengths (Å) and angles (°) for complexes **1** and **2**, Table S2: Continuous Shape Measurements for the ZnN_2O_4 coordination environment, Table S3: Continuous Shape Measurements for the CdN_2O_4 coordination environment.

Author Contributions: Conceptualization, A.R.-D.; methodology, A.A.G.-V. and E.E.-E.; software, J.C.; validation, J.M.D.-L. and S.R.; formal analysis, B.F. and S.R.; investigation, A.A.G.-V. and G.B.R.-R.; resources, B.F. and A.R.-D.; data curation, G.B.R.-R. and J.C.; writing—original draft preparation, E.E.-E.; writing—review and editing, J.C. and A.R.-D.; visualization, B.F.; supervision, A.R.-D.; project administration, J.C. and A.R.-D.; funding acquisition J.C. and A.R.-D. All authors have read and agreed to the published version of the manuscript.

Funding: Financial support was given by Junta de Andalucía (Spain) (FQM-394), University of the Basque Country (GIU 17/13), Gobierno Vasco/Eusko Jaurlaritza (IT1005-16), and the Spanish Ministry of Science, Innovation and Universities (MCIU/AEI/FEDER, UE) (PGC2018-102052-A-C22, PGC2018-102052-B-C21). J.M.D.L. and G.B.R.R. acknowledge the FEDER/MCIU/AEI for their Ramón y Cajal (RYC-2016-21042) and Juan de la Cierva (JdC-2017) fellowships, respectively. S.R. acknowledges the Juan de la Cierva Incorporación Fellowship (grant agreement n°. IJC2019-038894-I). E.E.-E. is grateful to the Government of the Basque Country for the predoctoral fellowship. The authors thank for technical and human support provided by SGIker of UPV/EHU and European funding (ERDF and ESF).

Institutional Review Board Statement: Not applicable.

Informed Consent Statement: Not applicable.

Conflicts of Interest: The authors declare no conflict of interest.

References

1. McKinlay, A.C.; Morris, R.E.; Horcajada, P.; Férey, G.; Gref, R.; Couvreur, P.; Serre, C. BioMOFs: Metal-organic frameworks for biological and medical applications. *Angew. Chem. Int. Ed.* **2010**, *49*, 6260–6266. [CrossRef]
2. Tranchemontagne, D.J.; Tranchemontagne, J.L.; O'keeffe, M.; Yaghi, O.M. Secondary building units, nets and bonding in the chemistry of metal–organic frameworks. *Chem. Soc. Rev.* **2009**, *38*, 1257–1283. [CrossRef] [PubMed]
3. Zhang, Q.; Cui, Y.; Qian, G. Goal-directed design of metal–organic frameworks for liquid-phase adsorption and separation. *Coord. Chem. Rev.* **2019**, *378*, 310–332. [CrossRef]
4. Bünzli, J.-C.G. On the design of highly luminescent lanthanide complexes. *Coord. Chem. Rev.* **2015**, *293–294*, 19–47. [CrossRef]
5. Chen, C.T.; Suslick, K.S. One-dimensional coordination polymers: Applications to material science. *Coord. Chem. Rev.* **1993**, *128*, 293–322. [CrossRef]
6. Heravi, M.M.; Zadsirjan, V. Prescribed drugs containing nitrogen heterocycles: An overview. *RSC Adv.* **2020**, *10*, 44247–44311. [CrossRef]
7. Horton, D.A.; Bourne, G.T.; Smythe, M.L. The combinatorial synthesis of bicyclic privileged structures or privileged substructures. *Chem. Rev.* **2003**, *103*, 893–930. [CrossRef]
8. Zhang, S.G.; Liang, C.G.; Zhang, W.H. Recent advances in indazole-containing derivatives: Synthesis and biological perspectives. *Molecules* **2018**, *23*, 2783. [CrossRef]
9. Dong, J.; Zhang, Q.; Wang, Z.; Huang, G.; Li, S. Recent Advances in the Development of Indazole-based Anticancer Agents. *ChemMedChem* **2018**, *13*, 1490–1507. [CrossRef]
10. Büchel, G.E.; Kossatz, S.; Sadique, A.; Rapta, P.; Zalibera, M.; Bucinsky, L.; Komorovsky, S.; Telser, J.; Eppinger, J.; Reiner, T.; et al. Cis -Tetrachlorido-bis(indazole)osmium(IV) and its osmium(III) analogues: Paving the way towards the cis -isomer of the ruthenium anticancer drugs KP1019 and/or NKP1339. *Dalt. Trans.* **2017**, *46*, 11925–11941. [CrossRef]
11. Hawes, C.S.; Kruger, P.E. Discrete and polymeric Cu(ii) complexes featuring substituted indazole ligands: Their synthesis and structural chemistry. *Dalt. Trans.* **2014**, *43*, 16450–16458. [CrossRef]

12. Long, B.F.; Huang, Q.; Wang, S.L.; Mi, Y.; Wang, M.F.; Xiong, T.; Zhang, S.C.; Yin, X.H.; Hu, F.L. Five new cobalt(II) complexes based on indazole derivatives: Synthesis, DNA binding and molecular docking study. *J. Coord. Chem.* **2019**, *72*, 645–663. [CrossRef]
13. Furman, J.D.; Burwood, R.P.; Tang, M.; Mikhailovsky, A.A.; Cheetham, A.K. Understanding ligand-centred photoluminescence through flexibility and bonding of anthraquinone inorganic-organic frameworks. *J. Mater. Chem.* **2011**, *21*, 6595–6601. [CrossRef]
14. Yuan, W.Z.; Shen, X.Y.; Zhao, H.; Lam, J.W.Y.; Tang, L.; Lu, P.; Wang, C.; Liu, Y.; Wang, Z.; Zheng, Q.; et al. Crystallization-induced phosphorescence of pure organic luminogens at room temperature. *J. Phys. Chem. C* **2010**, *114*, 6090–6099. [CrossRef]
15. Cepeda, J.; Rodríguez-Diéguez, A. Tuning the luminescence performance of metal-organic frameworks based on d10metal ions: From an inherent versatile behaviour to their response to external stimuli. *CrystEngComm* **2016**, *18*, 8556–8573. [CrossRef]
16. Solov'ev, K.N.; Borisevich, E.A. Intramolecular heavy-atom effect in the photophysics of organic molecules. *Phys. Uspekhi* **2005**, *48*, 231–253. [CrossRef]
17. Seco, J.M.; Pérez-Yáñez, S.; Briones, D.; García, J.Á.; Cepeda, J.; Rodríguez-Diéguez, A. Combining Polycarboxylate and Bipyridyl-like Ligands in the Design of Luminescent Zinc and Cadmium Based Metal-Organic Frameworks. *Cryst. Growth Des.* **2017**, *17*, 3893–3906. [CrossRef]
18. Cepeda, J.; Pérez-Yáñez, S.; Rodríguez-Diéguez, A. Luminescent Zn/Cd-based MOFS, CPS and their applications. In *Advances in Materials Science Research*; Wythers, M.C., Ed.; Nova Science Publishers: Hauppauge, NY, USA, 2017; Volume 28, ISBN 9781536109054.
19. San Sebastian, E.; Rodríguez-Diéguez, A.; Seco, J.M.; Cepeda, J. Coordination Polymers with Intriguing Photoluminescence Behavior: The Promising Avenue for Greatest Long-Lasting Phosphors. *Eur. J. Inorg. Chem.* **2018**, 2155–2174. [CrossRef]
20. Pajuelo-Corral, O.; Rodríguez-Diéguez, A.; García, J.A.; San Sebastián, E.; Seco, J.M.; Cepeda, J. Chiral coordination polymers based on d10metals and 2-aminonicotinate with blue fluorescent/green phosphorescent anisotropic emissions. *Dalt. Trans.* **2018**, *47*, 8746–8754. [CrossRef]
21. Yang, Y.; Wang, K.Z.; Yan, D. Ultralong Persistent Room Temperature Phosphorescence of Metal Coordination Polymers Exhibiting Reversible pH-Responsive Emission. *ACS Appl. Mater. Interfaces* **2016**, *8*, 15489–15496. [CrossRef]
22. Liu, Q.; Wang, R.; Wang, S. Blue phosphorescent Zn(II) and orange phosphorescent Pt(II) complexes of 4,4′-diphenyl-6,6′-dimethyl-2,2′-bipyrimidine. *Dalt. Trans.* **2004**, *35*, 2073–2079. [CrossRef]
23. Barbieri, A.; Accorsi, G.; Armaroli, N. Luminescent complexes beyond the platinum group: The d10 avenue. *Chem. Commun.* **2008**, 2185–2193. [CrossRef]
24. Hu, Z.; Deibert, B.J.; Li, J. Luminescent metal-organic frameworks for chemical sensing and explosive detection. *Chem. Soc. Rev.* **2014**, *43*, 5815–5840. [CrossRef] [PubMed]
25. Leo, P.; Briones, D.; García, J.A.; Cepeda, J.; Orcajo, G.; Calleja, G.; Rodríguez-Diéguez, A.; Martínez, F. Strontium-Based MOFs Showing Dual Emission: Luminescence Thermometers and Toluene Sensors. *Inorg. Chem.* **2020**, *59*, 18432–18443. [CrossRef]
26. Liu, J.; Zhuang, Y.; Wang, L.; Zhou, T.; Hirosaki, N.; Xie, R.-J. Achieving Multicolor Long-Lived Luminescence in Dye-Encapsulated Metal–Organic Frameworks and Its Application to Anticounterfeiting Stamps. *ACS Appl. Mater. Interfaces* **2018**, *10*, 1802–1809. [CrossRef]
27. Sun, Y.-X.; Sun, W.-Y. Zinc(ii)– and cadmium(ii)–organic frameworks with 1-imidazole-containing and 1-imidazole-carboxylate ligands. *CrystEngComm* **2015**, *17*, 4045–4063. [CrossRef]
28. Alvarez, S.; Avnir, D.; Llunell, M.; Pinsky, M. Continuous symmetry maps and shape classification. The case of six-coordinated metal compounds. *New J. Chem.* **2002**, *26*, 996–1009. [CrossRef]
29. Volkringer, C.; Loiseau, T.; Guillou, N.; Fèrey, G.; Haouas, M.; Taulelle, F.; Audebrand, N.; Margiolaki, I.; Popov, D.; Burghammer, M.; et al. Structural transitions and flexibility during dehydration—Rehydration process in the MOF-type aluminum pyromellitate A12(OH)2[C1008H2](MIL-118). *Cryst. Growth Des.* **2009**, *9*, 2927–2936. [CrossRef]
30. Boča, M.; Baran, P.; Boča, R.; Fuess, H.; Kickelbick, G.; Linert, W.; Renz, F.; Svoboda, I. Selective imidazolidine ring opening during complex formation of iron(III), copper(II), and zinc(II) with a multidentate ligand obtained from 2-pyridinecarboxaldehyde N-oxide and triethylenetetramine. *Inorg. Chem.* **2000**, *39*, 3205–3212. [CrossRef]
31. Xiao, H.; Li, X.B.; Qin, G.F.; Xia, Y.; Zhou, G. Directed assembly of cobalt(II) 1-H-indazole-3-carboxylic acid coordination networks by bipyridine and its derivatives: Structural versatility, electrochemical properties, and antifungal activity. *J. Iran. Chem. Soc.* **2016**, *13*, 793–802. [CrossRef]
32. Ohno, K.; Okimura, M.; Akai, N.; Katsumoto, Y. The effect of cooperative hydrogen bonding on the OH stretching-band shift for water clusters studied by matrix-isolation infrared spectroscopy and density functional theory. *Phys. Chem. Chem. Phys.* **2005**, *7*, 3005–3014. [CrossRef] [PubMed]
33. Salinas-Castillo, A.; Calahorro, A.J.; Briones, D.; Fairen-Jiménez, D.; Gándara, F.; Mendicute-Fierro, C.; Seco, J.M.; Pérez-Mendoza, M.; Fernández, B.; Rodríguez-Diéguez, A. 2D-cadmium MOF and gismondine-like zinc coordination network based on the N-(2-tetrazolethyl)-4′-glycine linker. *New J. Chem.* **2015**, *39*, 3982–3986. [CrossRef]
34. Pamei, M.; Puzari, A. Luminescent transition metal–organic frameworks: An emerging sensor for detecting biologically essential metal ions. *Nano-Struct. Nano-Objects* **2019**, *19*, 100364. [CrossRef]
35. García-Valdivia, A.A.; Zabala-Lekuona, A.; Ramírez-Rodríguez, G.B.; Delgado-López, J.M.; Fernández, B.; Cepeda, J.; Rodríguez-Diéguez, A. 2D-Coordination polymers based on 1H-indazole-4-carboxylic acid and transition metal ions: Magnetic{,} luminescence and biological properties. *CrystEngComm* **2020**, *22*, 5086–5095. [CrossRef]

36. García-Valdivia, A.A.; Pérez-Mendoza, M.; Choquesillo-Lazarte, D.; Cepeda, J.; Fernández, B.; Souto, M.; González-Tejero, M.; García, J.A.; Espallargas, G.M.; Rodríguez-Diéguez, A. Interpenetrated Luminescent Metal–Organic Frameworks based on 1H-Indazole-5-carboxylic Acid. *Cryst. Growth Des.* **2020**, *20*, 4550–4560. [CrossRef]
37. Dobretsov, G.E.; Syrejschikova, T.I.; Smolina, N.V. On mechanisms of fluorescence quenching by water. *Biophysics* **2014**, *59*, 183–188. [CrossRef]
38. Bruker Apex2. *B.A.I. Bruker Apex2*; Bruker AXS Inc.: Madison, WI, USA, 2004.
39. Sheldrick, G.M. SADABS 1996, Program for Empirical Adsorption Correction. Available online: https://cmacd.myweb.cs.uwindsor.ca/Teaching/553-class/sadabs.pdf (accessed on 16 January 2021).
40. Sheldrick, G.M. SHELXT—Integrated space-group and crystal-structure determination. *Acta Crystallogr. Sect. A Found. Crystallogr.* **2015**, *71*, 3–8. [CrossRef]

inorganics

MDPI

Article

Synthesis and Thermochromic Luminescence of Ag(I) Complexes Based on 4,6-Bis(diphenylphosphino)-Pyrimidine

Alexander V. Artem'ev *, Maria P. Davydova, Alexey S. Berezin and Denis G. Samsonenko

Nikolaev Institute of Inorganic Chemistry, Siberian Branch of Russian Academy of Sciences, 3, Acad. Lavrentiev Ave., 630090 Novosibirsk, Russia; m_davydova@mail.ru (M.P.D.); berezin@niic.nsc.ru (A.S.B.); denis@niic.nsc.ru (D.G.S.)

* Correspondence: chemisufarm@yandex.ru

Received: 6 August 2020; Accepted: 24 August 2020; Published: 26 August 2020

Abstract: Two Ag(I)-based metal-organic compounds have been synthesized exploiting 4,6-bis (diphenylphosphino)pyrimidine (L). The reaction of this ligand with $AgNO_3$ and $AgBF_4$ in acetonitrile produces dinuclear complex, $[Ag_2L_2(MeCN)_2(NO_3)_2]$ (**1**) and 1D coordination polymer, $[Ag_2L(MeCN)_3]_n(BF_4)_{2n}$ (**2**), respectively. In complex **1**, μ_2-P,P′-bridging coordination pattern of the ligand L is observed, whereas its μ_4-P,N,N′,P′-coordination mode appears in **2**. Both compounds exhibit pronounced thermochromic luminescence expressed by reversible changing of the emission chromaticity from a yellow at 300 K to an orange at 77 K. At room temperature, the emission lifetimes of **1** and **2** are 15.5 and 9.4 µs, the quantum efficiency being 18 and 56%, respectively. On account of temperature-dependent experimental data, the phenomenon was tentatively ascribed to alteration of the emission nature from thermally activated delayed fluorescence at 300 K to phosphoresce at 77 K.

Keywords: Ag(I) complexes; metal-organic coordination polymers; luminescence; thermally activated delayed fluorescence; phosphorescence; pyrimidylphosphines

1. Introduction

Recently, Ag(I)-based metal-organic compounds have attracted increased attention as promising antibacterial agents [1,2], luminescent sensorics [3,4], and potential emitters for lighting application [5–7]. The rich coordination abilities of Ag(I) ion coupled with easy accessibility of the organic ligands provide ample opportunities for the design of most diverse coordination architectures covering both simple molecules and sophisticated metal-organic frameworks [8–13]. The charge balancing anions (NO_3^-, OAc^-, OTf^-, ClO_4^-, BF_4^-, PF_6^-, halides, etc.,) can also determine the structure of products, which are self-assembled via the reactions of Ag(I) precursors with organic ligands [14–16]. Remarkably, such giant structural diversity of the Ag(I) compounds provides various tools for tuning their functional properties. For instance, they can be regulated by adjusting the electronic properties of organic ligands. Not the least, of the factors are supramolecular interactions of Ag···Ag, Ag···π, and Ag···X kind, which also can influence the properties of the Ag(I) compounds, e.g., luminescence [17–21].

Concerning luminescence of silver(I) metal-organic compounds, it was almost neglected for a long time, probably because of the preconception on photosensitivity of this class in principle. It was not until fairly recently the Ag(I) complexes were recognized as promising emitters, which can exhibit enhanced quantum efficiency coupled with short decay time [22–27]. Note that the luminescent properties of Ag(I) complexes primarily depend on the structure of the ligand environment. Compared to Cu(I) complexes showing a metal-to-ligand charge transfer (MLCT) luminescence [28,29], the Ag(I) analogues commonly emit metal-perturbed ligand-centered fluorescence [30–32], ligand-centered phosphorescence [33], or dual emission [34]. This stems from the fact that the MLCT excitation is hampered for Ag(I) complexes because of the higher ionization potential of Ag^+ ion in comparison with the Cu^+ one [35]. This restriction, however, can be overcome using highly electron-donating ligands (e.g., phosphines) coupled with π-acceptors (chelating diimines, azines, etc.,). On account of highly electron-donating ligands, the silver d^{10} orbitals begin to contribute to HOMO and near-HOMO, while the π-acceptors facilitate charge transfer from the metal. As a result, (M + L')LCT excited states can be generated, thereby inducing thermally activated delayed fluorescence (TADF) in Ag(I) compounds [22–27,36–42]. In the context of OLED application, it is relevant to note that the (M + L')LCT emission of Ag(I) benefits over that of Cu(I) analogues since the former (*i*) is shorter in the lifetimes [36,41,42], and (*ii*) generally appears in the higher energy domain [36,37,39,43–45]. On the whole, Ag(I) compounds that emit apparent luminescence at ambient temperature, especially, that of TADF nature, are still rare.

Herein, we report on the synthesis and investigation of two Ag(I) compounds derived from 4,6-bis(diphenylphosphino)pyrimidine and $AgNO_3$ or $AgBF_4$. Both compounds manifest pronounced thermochromic luminescence, which appears as a reversible changing of the emission color from yellow at 300 K to orange/red at 77 K. It should be noted in this regard that the Ag(I) complexes endowed with such property are very limited in number [13,46–49].

2. Results and Discussion

Synthesis and Characterization

The starting ligand, 4,6-bis(diphenylphosphino)pyrimidine (L), has been synthesized by the treatment of 2,6-dichloropyrimidine with 2 equiv. of lithium diphenylphosphide [50]. The ligand has been tested in the reaction with $AgNO_3$ and $AgBF_4$ in acetonitrile using different reactant's molar ratios. It has been revealed that the interaction of $AgNO_3$ with ligand L in a 1:1 molar ratio affords dinuclear complex $[Ag_2L_2(MeCN)_2(NO_3)_2]$ (**1**) isolated as solvate **1**·MeCN (Scheme 1). Meanwhile, using $AgBF_4$ under similar conditions, the reaction provides cationic 1D coordination polymer (CP), $[Ag_2L(MeCN)_3]_n(BF_4)_{2n}$ (**2**) that also crystallizes as solvate **2**·MeCN (Scheme 1). The preparative yields of products **1** and **2** are 69 and 91%, respectively.

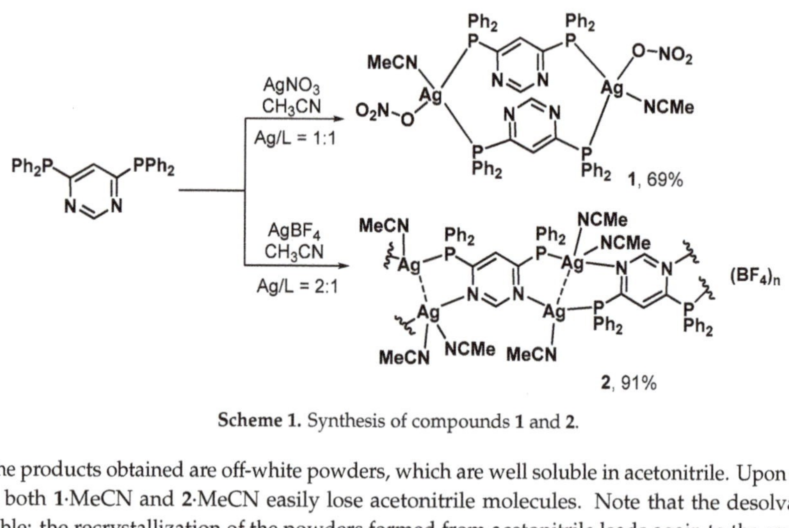

Scheme 1. Synthesis of compounds **1** and **2**.

The products obtained are off-white powders, which are well soluble in acetonitrile. Upon storage on air, both **1**·MeCN and **2**·MeCN easily lose acetonitrile molecules. Note that the desolvation is reversible: the recrystallization of the powders formed from acetonitrile leads again to the crystals of the above solvates. Both compounds have been characterized by single crystal X-ray diffractometry (sc-XRD), FT-IR, and UV-Vis abortion spectroscopy.

Complex **1**·MeCN crystallizes in the monoclinic $P2_1/n$ space group with one half molecule per asymmetric unit. As seen from Figure 1, the scaffold of **1** is formed by two Ag atoms bridged by two ligands L through phosphorus atoms so that the pyrimidine rings become coplanar. The latter are sandwiched in a "head-to-tail" manner with the distance between the average planes being 3.861 Å. The O atom of the NO_3 group and acetonitrile N atom complete the coordination sphere of Ag1 and Ag2 atoms to the distorted {Ag@P_2ON} tetrahedron. The bond lengths around metal atoms are comparable with those of most related Ag(I) complexes [51–53]. In the crystal, molecules of **1** are associated together and with MeCN molecules via weak C–H···O and C–H···C contacts forming 3D supramolecular structure.

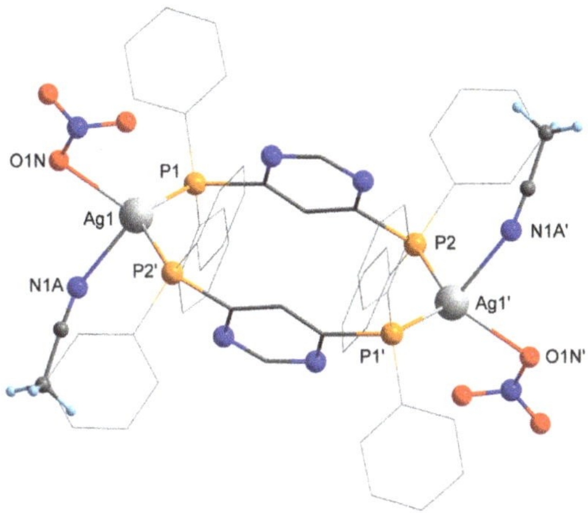

Figure 1. Molecular structure of **1**·MeCN. The aromatic H atoms and solvate molecules are omitted. Selected bond lengths (Å) and angles (°): Ag1–P1 2.4224(13), Ag1–P2′ 2.4461(13), Ag1–O1N 2.273(9), Ag1–N1A 2.493(8); P1–Ag1–P2′ 118.48(4), P1–Ag1–O1N 123.4(4), P1–Ag1–N1A 107.0(3), P2′–Ag1–O1N 110.9(3), P2′–Ag1–N1A 96.8(3), O1N–Ag1–P1 134.1(3), O1N–Ag1–N1A 79.8(4). Symmetry code (′): $1 - x, 1 - y, 1 - z$.

CP **2**·MeCN crystallizes in the orthorhombic $P2_12_12_1$ space group, and its crystals contain zig-zag chains propagating along *a* axis. The chains are built up from alternating ligand molecules and $[Ag_2(MeCN)_3]$ units, which are linked via Ag–N and Ag–P bonds (Figure 2). The metal atoms of the $[Ag_2(MeCN)_3]$ units are bridged by two P,N-faces of the adjacent ligands L in a "head-to-tail" fashion. The formed eight-membered cycles feature short Ag···Ag contact of 3.3352(4) Å that is consistent with twice van der Waals radius of Ag atom (3.44 Å [54]). Both Ag1 and Ag2 atoms are ligated by one MeCN molecule ($d_{Ag\text{-}N} \approx 2.33$ Å); besides, a second MeCN molecule is weakly associated ($d_{Ag\text{-}N} \approx 2.65$ Å) with Ag1 atom. As a result, Ag1 atom adopts a T-shaped {Ag@N_2P} geometry, while Ag2 center has a distorted see-saw {Ag@N_3P} environment. The Ag–N_L and Ag–P distances are nearly the same as those in **1**·MeCN. The non-coordinated tetrafluoroborate anions and MeCN solvate molecules are associated with the $[Ag_2L(MeCN)_3]_n$ chains by means of van der Waals contacts such as C–H···F, C–H···N, Ag···F, and C···F.

FT-IR spectra of solid **1** and **2** are in agreement with sc-XRD data, showing characteristic vibrations of the ligand L along with stretching vibrations of the counter-ions (Figure S1). The N–O and B–F stretchings of the NO_3^- and BF_4^- groups appear as strong bands at 1380–1417 and 950–1200 cm^{-1}, correspondingly. The solid state UV-Vis spectra of **1** and **2** (plotted as Kubelka-Munk function, Figure S2) display broad bands expanding from the far-UV edge and falling close at about 400 and 440 nm, respectively (Figure S2). Each absorption band has two pronounced shoulders. The high-energy (HE) shoulders, with $\lambda_{max} \approx 280$ nm, are nearly the same for both compounds, while the low-energy (LE) ones maximize at ≈350 nm for **1** and at ≈390 nm for **2**. Considering the literature data [38], the HE absorption band can be attributed to intraligand π–π * and n–π * transitions. The LE band is likely associated with promotions of MLCT kind, which is typical for emissive Ag(I) complexes [22–27,36–40].

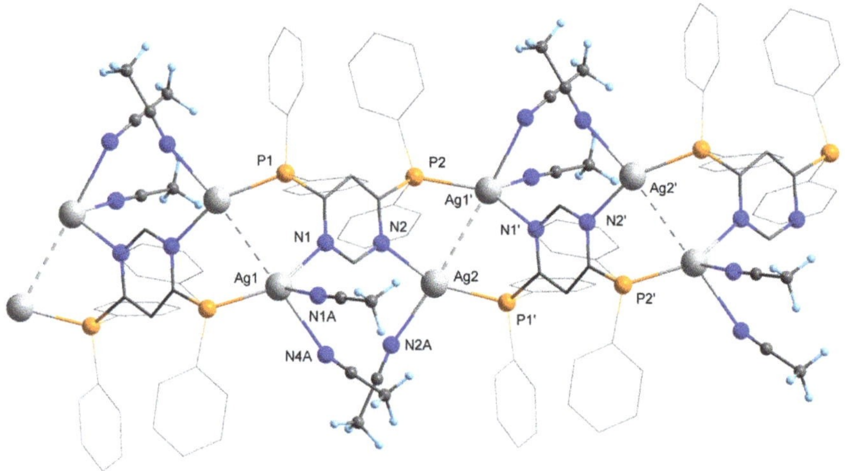

Figure 2. A fragment of the 1D chain of **2**·MeCN. The aromatic H atoms, $[BF_4]^-$ counterions and solvate molecules are omitted. Selected bond lengths (Å) and angles (°): Ag1′···Ag2 3.3352(4), Ag1′–P2 2.3943(9), Ag1–N1 2.309(3), Ag1–N1A 2.320(4), Ag1–N4A 2.655(4), Ag2–P1′ 2.3622(9), Ag2–N2 2.242(3), Ag2–N2A 2.351(4); P2–Ag1′–Ag2 69.98(2), N1–Ag1–N1A 98.43(11), N1–Ag1–N4A 87.69(10), N1A–Ag1–N4A 81.28(13), P1′–Ag2–Ag1′ 72.95(2), N2–Ag2–Ag1′ 86.93(7), N2–Ag2–P1′ 152.18(7), N2–Ag2–N2A 92.88(12), N2A–Ag2–Ag1′ 165.92(12), N2A–Ag2–P1′ 111.31(9). Symmetry code ('): $0.5 + x, 1.5 - y, 1 - z$.

When UV-irradiated, solid compounds **1** and **2** exhibit yellow luminescence at ambient temperature. Upon cooling down to liquid nitrogen temperature, the luminescence strongly enhances, and its color changes to red-orange (for **1**) or red (for **2**) (Figures 3d and 4d). The thermochromic luminescence found appears to be reversible: warming the samples to 300 K recovers the initial emission chromaticity. Inspirited by these noticeable findings, we have studied the emission properties of the titled compounds at 77–300 K range. Temperature-dependent emission and excitation spectra of **1** and **2** are plotted in Figures 3 and 4, and the corresponding photophysical data are summarized in Table 1. As seen from the graphs, the emission spectra of **1** and **2** contain a broad band maximized at about 550 and 580 nm, accordingly. The corresponding emission colors on the CIE chromaticity diagram are consistent with those observed by the naked eye. The associated emission lifetimes (τ_{obs}) of **1** and **2** measured at 300 K are 15.5 and 9.4 µs, and the photoluminescence quantum yields (Φ_{PL}) are 18 and 56%, respectively. The excitation profiles of **1** and **2** are presented by smooth bands that fall close at about 420 and 440 nm (Figures 3b and 4b). The excitation curves, therefore, resemble the absorption patterns (Figure S2). Note that the compounds studied do not possess excitation-dependent properties, which are quite common for Ag(I) complexes [55–57]. When the temperature is gradually lowered to 77 K, the emission bands of **1** and **2** are red-shifted by 30 and 22 nm (Figures 3a and 4a), thereby changing the emission color to red-orange and red, respectively (Figure 3c,d). Simultaneously, the lifetimes rise to 3970 µs (**1**) and 300 µs (**2**).

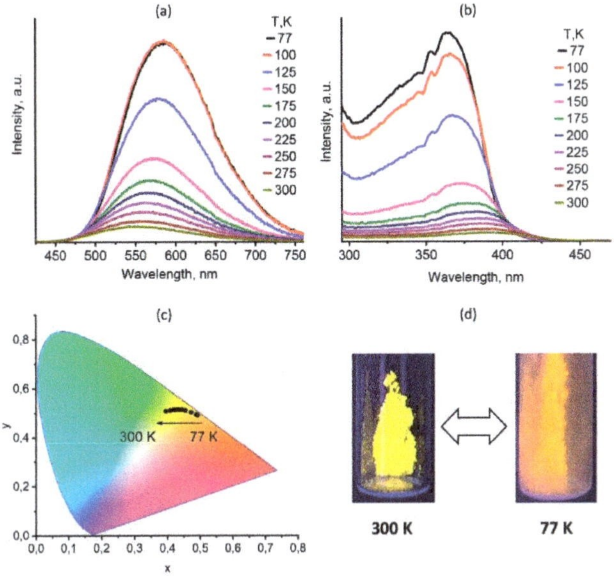

Figure 3. (**a**) Temperature-dependent emission spectra of **1** (λ_{ex} = 385 nm); (**b**) temperature-dependent excitation spectra of **1** (λ_{em} = 580 nm); (**c**) temperature dependence of the emission chromaticity of **1** (λ_{ex} = 385 nm); (**d**) emission color of sample **1** at 300 and 77 K.

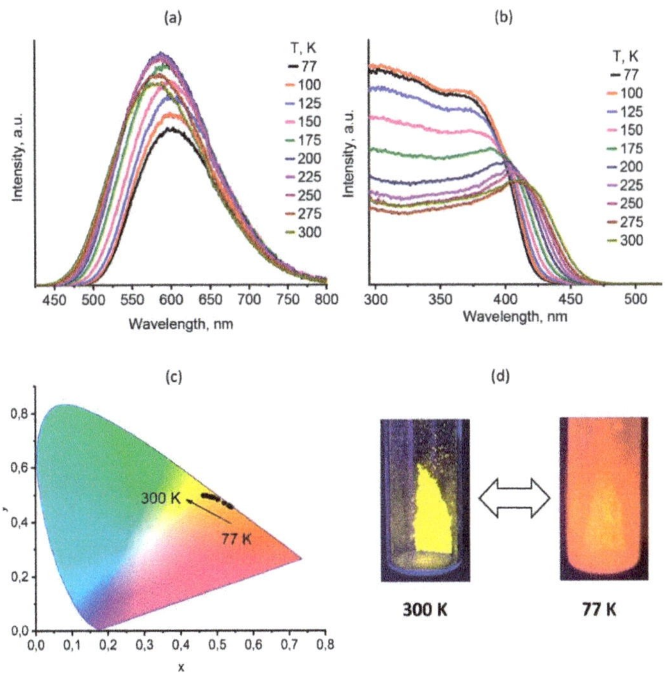

Figure 4. (**a**) Temperature-dependent emission spectra of **2** (λ_{ex} = 410 nm); (**b**) temperature-dependent excitation spectra of **2** (λ_{em} = 580 nm); (**c**) temperature dependence of the emission chromaticity of **2** (λ_{ex} = 385 nm); (**d**) emission color of sample **2** at 300 and 77 K.

Table 1. Photophysical data for solid **1** and **2**.

	1	**2dd**
λ_{max} (300 K) (nm)	550	580
Φ_{PL} (300 K) (%)	18 [a]	56 [b]
τ (300 K) (μs)	15.5	9.4
λ_{max} (77 K) (nm)	580	602
τ (77 K) (μs)	3970	300

[a] λ_{ex} = 365 nm; [b] λ_{ex} = 410 nm.

Taken together, these observations suggest that TADF is likely responsible for the room temperature emission of **1** and **2**. The temperature dependence of the lifetimes, τ_{obs} (T), measured in 77–300 K window, supports this suggestion, following the equation intended for the TADF model [58]:

$$\tau_{obs}(T) = (3 + \exp\left(-\frac{\Delta E_{ST}}{k_B T}\right))/(\frac{3}{\tau_T} + \frac{1}{\tau_S}\exp(-\frac{\Delta E_{ST}}{k_B T})) \quad (1)$$

wherein τ_S and τ_T are the lifetimes of prompt fluorescence and phosphorescence, respectively, ΔE_{ST} is the energy gap between the respective excited states (S_1 and T_1), and k_B is the Boltzmann constant. Applying this equation for fitting the datasets of Figure 5, the following values have been roughly estimated for **1**: $\Delta E(S_1{-}T_1)$ = 750 cm^{-1}, fluorescence lifetime τ (S_1) = 400 ns, and phosphorescence lifetime τ (T_1) = 4000 μs. Analogously, the following values have been estimated for **2**: $\Delta E(S_1{-}T_1)$ = 1000 cm^{-1}, τ (S_1) = 35 ns, and τ (T_1) = 300 μs. It should be emphasized that the given $\Delta E(S_1{-}T_1)$ values are purely evaluative because the Eq. 1 can be applied correctly when the emission quantum yields do not change much than the investigated temperature range [58]. Nevertheless, the estimated $\Delta E(S_1{-}T_1)$ magnitudes agree well with the common values for TADF-emitting Ag(I) and Cu(I) complexes, thus allowing to assume manifestation of TADF by **1** and **2**.

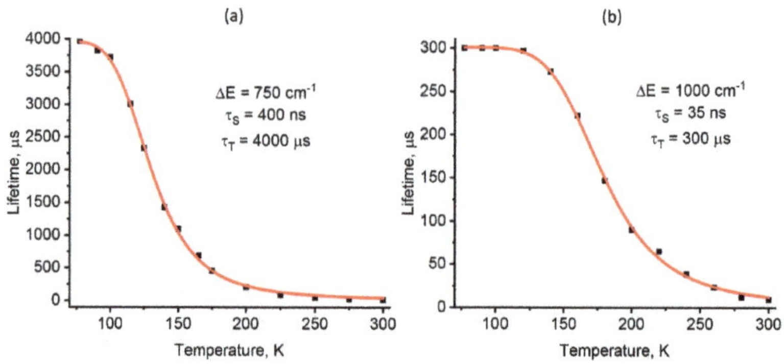

Figure 5. (**a**) Temperature dependence of the emission lifetimes for **1** (λ_{ex} = 385 nm, λ_{em} = 550 nm), and (**b**) for **2** (λ_{ex} = 410 nm, λ_{em} = 580 nm).

As seen from Figure 5, the luminescence of both compounds at ambient temperature represents TADF because the τ_{obs} (T) curves attain the high-temperature plateau. The pure phosphorescence begins to appear when the τ_{obs} (T) curve reaches the low-temperature plateau. In the case of **2**, it occurs below 120 K, while complex **1**, possessing a narrower $\Delta E(S_1{-}T_1)$ gap, emits pure phosphorescence at 77 K and below. On the drop-down range of the τ_{obs} (T) curves, the contribution of the TADF gradually increases up to ~100% at 300 K, owing to the thermal population of the higher-lying S_1 state from the T_1 state. As a result, the emission energy shifts in the blue region upon warming from 77 to 300 K, resulting in thermochromic luminescence of **1** and **2**. Considering the literature data on the related Ag(I) complexes exhibiting TADF [22–27,36–42], we believe that the S_1 and T_1

excited states of **1** and **2** are of MLCT or (M + L′)LCT nature. The DFT computations performed on complex **1** support this suggestion revealing that HOMO and near-HOMOs are contributed by silver d-orbitals and p-orbitals of nitrate oxygen atoms, while LUMO and near-LUMOs are pure pyrimidine π-orbitals (Figure S10). Note the previously reported Ag(I) complexes feature a similar HOMO/LUMO distribution pattern [22–27,36–42]. Thus, it can be assumed that the emissive excited state of **1** has a ^{1,3}(M + L′)LCT character (L′ = NO_3, and L = pyrimidyldiphosphine).

3. Materials and Methods

3.1. General

Synthesis of the ligand L was performed under argon atmosphere, while the compounds **1** and **2** were prepared under ambient conditions. $AgNO_3$ (≥99.9%, Aldrich, St. Louis, Missouri, MO, USA), $AgBF_4$ (≥99.9%, Aldrich), *n*-BuLi (2.5 M in hexanes, Aldrich), 4,6-dichloropyrimidine (97%, Aldrich), and diphenylphosphine (98%, Aldrich) were used as purchased. Prior to use, commercial tetrahydrofuran (THF, anhydrous, ≥99.9%, Aldrich) was purified by distillation over sodium/benzophenone under argon flow. Acetonitrile and dichloromethane were distilled over phosphorus pentoxide.

FT-IR spectra were measured on a Bruker Vertex 80 spectrometer (Bruker, Billerica, Massachusetts, MA, USA) at ambient temperature. The microanalyses were performed on a MICRO cube analyzer. ^{1}H, ^{13}C, and $^{31}P\{^{1}H\}$ NMR spectra were recorded using a Bruker AV-500 spectrometer at 500.13, 125.8 MHz and 202.46 MHz, respectively, with solvent peaks as reference. The $^{31}P\{^{1}H\}$ NMR shifts are expressed with respect to 85% H_3PO_4/D_2O as an external standard.

The microanalyses were performed on a MICRO cube analyzer Photoluminescence spectra were recorded on a Fluorolog 3 spectrometer (Horiba Jobin Yvon, Kyoto, Japan) with a cooled PC177CE-010 photon detection module equipped with an R2658 photomultiplier. The luminescence decays (Figures S5 and S6) were measured on the same instrument. The absolute values of PLQYs were recorded using a Fluorolog 3 Quanta-phi device (Horiba Jobin Yvon). The luminescence quantum yield at 77 K was obtained relative to the quantum yield of the same sample at 300 K. Independently, these relative quantum yields were calibrated by using the absolute PLQY values measured at 77 K. Temperature dependences of luminescence were carried out using Optistat DN optical cryostats (Oxford Instruments, Abingdon, UK).

The solid-state reflectance spectra were recorded on a Shimadzu UV-3101 spectrophotometer (Shimadzu Corporation, Kyoto, Japan). Samples were prepared by a thorough grinding of a mixture of a complex (*ca.* 2 mol %) with $BaSO_4$. The reflectance data were converted into a spectrum applying a Kubelka-Munk function using $BaSO_4$ as a standard.

DFT computations of **1** were performed using the hybrid B3LYP functional [59] combined with the def2TZVP basis sets [60]. The calculations were performed for single point geometry taken from X-ray coordinates for non-hydrogen atoms. The computations were proceeded using Gaussian 09 suite [61].

3.2. Synthesis of 4,6-Bis(diphenylphosphino)pyrimidine (L)

n-BuLi in hexanes (2.5 M, 15 mL) was added dropwise to a solution of diphenylphosphine (6.139 g, 0.033 mol) in absolute THF (40 mL) at −20 °C. The mixture was kept at −20 °C and stirred for 1 h. Then, at the same temperature, suspension of 4,6-dichloropyrimidine (2.384 g, 0.016 mol) in THF (10 mL) was added dropwise. The resulting mixture was warmed to 40 °C and stirred for 4 h. After that H_2O (50 mL) was added and the quenched mixture was extracted with CH_2Cl_2 (3 × 30 mL). The organic extracts were washed with H_2O (3 × 10 mL), dried with Na_2SO_4, and evaporated in vacuum. The crude product obtained was recrystallized from $MeOH/CH_2Cl_2$ (10:1, *v*/*v*) to give colorless crystals of L. Yield: 4.087 g (57%). ^{1}H NMR (500.13 MHz, $CDCl_3$) δ 9.20 (s, 1H, C^2–H in Pym), 7.35–7.31 (m, 20H in Ph), 6.69 (s, 1H, C^5–H in Pym). $^{13}C\{^{1}H\}$ NMR (126 MHz, $CDCl_3$) δ 174.4 (d, *J* = 7.0 Hz, C^4 and C^6 in Pym), 156.8 (t, *J* = 9.3 Hz, C^2 in Pym), 134.4 (s, *o*-Ph), 134.3 (s, *o*-Ph), 133.6 (d, *J* = 8.8 Hz, *i*-Ph), 132.1

(d, J = 9.6 Hz, C^5 in Pym), 129.7 (s, p-Ph), 128.9 (d, J = 7.8 Hz, m-Ph). $^{31}P\{^1H\}$ NMR δ (202.47 MHz, $CDCl_3$) −2.62. FT-IR (KBr, cm^{-1}): 424 (w), 434 (w), 444 (m), 463 (w), 484 (m), 500 (vs), 544 (w), 608 (w), 619 (w), 696 (vs), 743 (vs), 766 (m), 783 (w), 889 (w), 978 (w), 999 (m), 1026 (m), 1070 (w), 1097 (m), 1157 (w), 1184 (w), 1260 (s), 1310 (w), 1331 (w), 1435 (s), 1479 (s), 1535 (s), 1584 (w), 2984 (vw), 3049 (w), 3071 (vw).

3.3. *$[Ag_2L_2(NO_3)_2(MeCN)_2]$·MeCN (**1**·MeCN)*

A solution of L (50 mg, 0.11 mmol) and $AgNO_3$ (20 mg, 0.11 mmol) in CH_3CN (1 mL) was stirred at room temperature for 30 min. The precipitated white powder of the **1**·CH_3CN was centrifuged and dried in air. White powder. Yield: 50 mg (69%). Single crystals of **1**·CH_3CN were grown by vapor diffusion of diethyl ether into the CH_3CN solution for overnight. FT-IR (thin film, cm^{-1}): 474 (m), 505 (s), 692 (vs), 746 (s), 997 (w), 1028 (w), 1099 (m), 1265 (m), 1288 (s), 1385 (s), 1420 (s), 1437 (vs), 1481 (m), 1495 (m), 1539 (vs), 2251 (w), 2294 (vw), 2921 (w), 2936 (vw), 2994 (vw), 3056 (w). Anal. Calcd: $C_{56}H_{44}Ag_2N_6P_4O_6$ (1236.62) C, 54.4; H, 3.6; N, 6.8. Found: C, 54.4; H, 3.4; N, 6.7. Since the solvate **1**·CH_3CN quickly loses the molecules of acetonitrile upon storage, the microanalysis was calculated on $[Ag_2(L)_2(NO_3)_2]$.

3.4. *$[Ag_2L(MeCN)_3]_n(BF_4)_{2n}$·MeCN (**2**·MeCN)*

A solution of L (50 mg, 0.11 mmol) and $AgBF_4$ (43 mg, 0.22 mmol) in CH_3CN (1 mL) was stirred at room temperature for 30 min. To the resulting solution, diethyl ether (5 mL) was then added and the precipitate formed was centrifuged and dried in air. White powder. Yield: 100 mg (91%). Single crystals of **2**·CH_3CN were grown by vapor diffusion of diethyl ether into the CH_3CN solution for overnight. FT-IR (cm^{-1}): 478 (w), 507 (m), 519 (m), 692 (S), 746 (m), 997 (m), 1063 (vs), 1084 (vs), 1097 (vs), 1163 (w), 1184 (w), 1287 (w), 1308 (w), 1437 (s), 1454 (w), 1481 (w), 1497 (w), 1560 (m), 1634 (w), 2253 (w), 2298 (w), 2388 (w), 2971 (w), 3009 (vw), 3063 (w). Anal. Calcd: $C_{28}H_{22}Ag_2B_2F_8N_2P_2$ (837.78) C, 40.1; H, 2.6; N, 3.3. Found: C, 40.0; H, 2.6; N, 3.5. Since the solvate **2**·CH_3CN quickly loses the molecules of acetonitrile upon storage, the microanalysis was calculated on $[Ag_2L](BF_4)_2$.

3.5. *X-ray Crystallography*

Single crystals of **1**·MeCN and **2**·MeCN were grown by diffusion of diethyl ether vapors into a MeCN solutions at ambient temperature for overnight. The X-ray data and the details of the refinement are summarized in Table S1. Diffraction data were collected on an automated Agilent Xcalibur diffractometer equipped with an area AtlasS2 detector (graphite monochromator, λ (Mo Kα) = 0.71073 Å, ω-scans, Agilent, Santa Clara, California, CA, USA). Integration, absorption correction, and determination of unit cell parameters were performed using the CrysAlisPro program package [62]. The structures were solved by dual space algorithm (SHELXT [63]) and refined by the full-matrix least squares technique (SHELXL [64]) in the anisotropic approximation (except hydrogen atoms). Positions of hydrogen atoms of organic ligands were calculated geometrically and refined in the riding model.

The crystallographic data and details of the structure refinements are summarized in Table S1. CCDC 2020455 and 2020456 contain the supplementary crystallographic data for this paper. These data can be obtained free of charge from The Cambridge Crystallographic Data Center at http://www.ccdc.cam.ac.uk/data_request/cif.

4. Conclusions

Thus, two Ag(I) metal-organic compounds have been synthesized by the treatment of 4,6-bis(diphenylphosphino)pyrimidine (L) with $AgNO_3$ and $AgBF_4$ in acetonitrile. It has been revealed that the interaction with $AgNO_3$ results in neutral dinuclear complex, $[Ag_2L_2(MeCN)_2(NO_3)_2]$, while the reaction with $AgBF_4$ produces cationic 1D zig-zag polymer, $[Ag_2L(MeCN)_3]_n(BF_4)_{2n}$. The structure of the complex is built up from two Ag(I) ions bridged by the two ligands in a

μ_2-P,P′-manner. The 1D chains of the polymer are assembled by alternating ligand (μ_4-N,P,N′,P′) and [$Ag_2(MeCN)_3$] units, interconnected through Ag–N and Ag–P bonds. Both title compounds feature pronounced thermochromic luminescence, which appears as reversible yellow-to-orange changing of the emission color during the cooling-warming cycling (300–77 K). The detailed temperature-dependent photophysical study has shown that the ambient temperature photoluminescence of the above compounds may be tentatively ascribed to TADF. At 77 K, they certainly emit pure phosphorescence. The distinct thermochromic behavior of the complexes designed makes them promising materials for luminescent thermometry. From the fundamental viewpoint, the findings reported contribute to coordination chemistry and photophysics of Ag(I)-based metal-organic compounds.

Supplementary Materials: The following are available online at http://www.mdpi.com/2304-6740/8/9/46/s1, Figure S1: FT-IR spectra of **1** and **2** displayed in the fingerprint range; Figure S2: Solid state absorption spectra of **1** and **2** plotted as a Kubelka–Munk function; Figure S3: Temperature dependence of the integral intensity of the emission of **1** (λ_{ex} = 365 nm); Figure S4: Temperature dependences of the integral intensity of the emission of **2** recorded at λ_{ex} = 410 nm (*left*) and λ_{ex} = 365 nm (*right*); Figure S5: Emission decay profiles of **1** recorded at different temperatures (λ_{ex} = 385 nm, λ_{em} = 550 nm); Figure S6: Emission decay profiles of **2** recorded at different temperatures (λ_{ex} = 410 nm, λ_{em} = 580 nm); Figure S7: ^{1}H NMR spectrum of the ligand **L** ($CDCl_3$); Figure S8: J-modulated ^{13}C NMR spectrum of the ligand **L** ($CDCl_3$); Figure S9: ^{31}P{^{1}H} NMR spectrum of the ligand **L** ($CDCl_3$); Figure S10: Four lowest unoccupied and 4 highest occupied MOs (iso-value = 0.045) for the S_0 state of the complex **1** calculated at the B3LYP/def2TZVP level; Table S1: X-Ray crystallographic data for **1**·CH_3CN and **2**·CH_3CN; the CIF and the checkCIF output files are included in the Supplementary Materials.

Author Contributions: Project conceptualization, administration, supervision, writing—review and editing, and funding acquisition, A.V.A.; investigation, writing, data curation, visualization, M.P.D. and A.S.B.; crystallography, D.G.S. All authors have read and agreed to the published version of the manuscript.

Funding: This work was supported by Russian Science Foundation (Project 18-73-10086).

Acknowledgments: We thank Evgeniya Doronina (A.E. Favorsky Irkutsk Institute of Chemistry, Irkutsk, Russian Federation) for help with the DFT calculations.

Conflicts of Interest: The authors declare no conflict of interest.

References

1. Medici, S.; Peana, M.; Crisponi, G.; Nurchi, V.M.; Lachowicz, J.I.; Remelli, M.; Zoroddu, M.A. Silver coordination compounds: A new horizon in medicine. *Coord. Chem. Rev.* **2016**, *327–328*, 349–359. [CrossRef]
2. Kulovi, S.; Dalbera, S.; Das, S.; Zangrando, E.; Puschmann, H.; Dalai, S. New silver(I) coordination polymers with hetero donor ligands: Synthesis, structure, luminescence study and photo-catalytic behavior. *ChemistrySelect* **2017**, *2*, 9029–9036. [CrossRef]
3. Rawashdeh-Omary, M.A.; Rashdan, M.D.; Dharanipathi, S.; Elbjeirami, O.; Ramesh, P.; Rasika Dias, H.V. On/off luminescence vapochromic selective sensing of benzene and its methylated derivatives by a trinuclear silver(I) pyrazolate sensor. *Chem. Commun.* **2011**, *47*, 1160–1162. [CrossRef] [PubMed]
4. Wang, Y.; Shi, Y.; Zou, X.; He, Y.; Wang, X. Pyridylphosphine supported Ag(I) and Cu(I) complexes for detection of alcohols and nitriles via structural transformations from 1D to 0D. *CrystEngComm* **2019**, *21*, 5595–5601. [CrossRef]
5. Fresta, E.; Carbonell-Vilar, J.M.; Yu, J.; Armentano, D.; Cano, J.; Viciano-Chumillas, M.; Costa, R.D. Deciphering the electroluminescence behavior of silver(I)-complexes in light-emitting electrochemical cells: Limitations and solutions toward highly stable devices. *Adv. Funct. Mater.* **2019**, *29*, 1901797. [CrossRef]
6. Moudam, O.; Tsipis, A.C.; Kommanaboyina, S.; Horton, P.N.; Coles, S.J. First light-emitting electrochemical cell with [Ag (I)(N^N)(P^P)] type complex. *RSC Adv.* **2015**, *5*, 95047–95053. [CrossRef]
7. Lu, Z.; Cheng, Y.; Fan, W.; Yang, S.; Liu, X.; Qin, Y.; Zhao, R.; Zheng, L.; Zhang, H. A stable silver metallacage with solvatochromic and mechanochromic behavior for white LED fabrication. *Chem. Commun.* **2019**, *55*, 8474–8477. [CrossRef]
8. Khlobystov, A.N.; Blake, A.J.; Champness, N.R.; Lemenovskii, D.A.; Majouga, A.G.; Zyk, N.V.; Schröder, M. Supramolecular design of one-dimensional coordination polymers based on silver(I) complexes of aromatic nitrogen-donor ligands. *Coord. Chem. Rev.* **2001**, *222*, 155–192. [CrossRef]

9. Sharma, S.; Chakrahari, K.K.; Saillard, J.-Y.; Liu, C.W. Structurally precise dichalcogenolate-protected copper and silver superatomic nanoclusters and their alloys. *Acc. Chem. Res.* **2018**, *51*, 2475–2483. [CrossRef]
10. Huang, R.; Wei, Y.; Dong, X.; Wu, X.-H.; Du, C.-X.; Zang, S.-Q.; Mak, T.C.W. Hypersensitive dual-function luminescence switching of a silver-chalcogenolate cluster-based metal–organic framework. *Nat. Chem.* **2017**, *9*, 689–697. [CrossRef]
11. Jin, G.-X.; Zhu, G.-Y.; Sun, Y.-Y.; Shi, Q.-X.; Liang, L.-P.; Wang, H.-Y.; Wu, X.-W.; Ma, J.-P. $[Ag–Ag]^{2+}$ unit-encapsulated trimetallic cages: One-pot syntheses and modulation of argentophilic interactions by the uncoordinated substituents. *Inorg. Chem.* **2019**, *58*, 2916–2920. [CrossRef]
12. Chupina, A.V.; Mukhacheva, A.A.; Abramov, P.A.; Sokolov, M.N. Complexation and isomerization of $[\beta\text{-}Mo_8O_{26}]^{4-}$ in the presence of Ag^+ and DMF. *J. Struct. Chem.* **2020**, *61*, 299–308. [CrossRef]
13. Shmakova, A.A.; Berezin, A.S.; Abramov, P.A.; Sokolov, M.N. Self-assembly of $Ag^+/[PW_{11}NbO_{40}]^{4-}$ complexes in nonaqueous solutions. *Inorg. Chem.* **2020**, *59*, 1853–1862. [CrossRef] [PubMed]
14. Wu, J.-Y.; Chao, T.-C.; Zhong, M.-S. Influence of counteranions on the structural modulation of silver–di(3-pyridylmethyl)amine coordination polymers. *Cryst. Growth Des.* **2013**, *13*, 2953–2964. [CrossRef]
15. Liu, F.-J.; Sun, D.; Hao, H.-J.; Huang, R.-B.; Zheng, L.-S. Anion-controlled assembly of silver(I)/aminobenzonitrile compounds: Syntheses, crystal structures, and photoluminescence properties. *Cryst. Growth Des.* **2012**, *12*, 354–361. [CrossRef]
16. Seward, C.; Chan, J.; Song, D.; Wang, S. Anion dependent structures of luminescent silver(I) complexes. *Inorg. Chem.* **2003**, *42*, 1112–1120. [CrossRef] [PubMed]
17. Che, C.-M.; Tse, M.-C.; Chan, M.C.W.; Cheung, K.-K.; Phillips, D.L.; Leung, K.-H. Spectroscopic evidence for argentophilicity in structurally characterized luminescent binuclear silver(I) complexes. *J. Am. Chem. Soc.* **2000**, *122*, 2464–2468. [CrossRef]
18. Feazell, R.P.; Carson, C.E.; Klausmeyer, K.K. Variability in the structures of luminescent [2-(aminomethyl)pyridine] silver(I) complexes: Effect of ligand ratio, anion, hydrogen bonding, and π-stacking. *Eur. J. Inorg. Chem.* **2005**, *2005*, 3287–3297. [CrossRef]
19. Matos, C.R.M.O.; Monteiro, F.G.A.; Miranda, F.S.; Pinheiro, C.B.; Bond, A.D.; Ronconi, C.M. Tuning Photoluminescent properties of silver(I)-based coordination networks through their supramolecular interactions. *Cryst. Growth Des.* **2017**, *17*, 5965–5974. [CrossRef]
20. Zhan, S.-Z.; Song, H.-Q.; Guo, L.-J.; Sun, R.W.-Y.; Li, D. Structure- and temperature-dependent luminescence properties of threefold interpenetrated networks: Coordination polymers based on dinuclear gridlike silver(I) units. *Eur. J. Inorg. Chem.* **2017**, *2017*, 5127–5133. [CrossRef]
21. Liu, S.Q.; Kuroda-Sowa, T.; Konaka, H.; Suenaga, Y.; Maekawa, M.; Mizutani, T.; Ning, G.L.; Munakata, M. Silver(I) coordination polymers of fluorescent oligo (phenylenevinylene) with π–π stackings: luminescence and conductivity. *Inorg. Chem.* **2005**, *44*, 1031–1036. [CrossRef] [PubMed]
22. Jia, J.-H.; Liang, D.; Yu, R.; Chen, X.-L.; Meng, L.; Chang, J.-F.; Liao, J.-Z.; Yang, M.; Li, X.-N.; Lu, C.-Z. Coordination-induced thermally activated delayed fluorescence: From non-TADF donor–acceptor-type ligand to TADF-active Ag-based complexes. *Chem. Mater.* **2020**, *32*, 620–629. [CrossRef]
23. Chen, J.; Teng, T.; Kang, L.; Chen, X.-L.; Wu, X.-Y.; Yu, R.; Lu, C.-Z. Highly efficient thermally activated delayed fluorescence in dinuclear Ag(I) complexes with a bis-bidentate tetraphosphane bridging ligand. *Inorg. Chem.* **2016**, *55*, 9528–9536. [CrossRef] [PubMed]
24. Gan, X.-M.; Yu, R.; Chen, X.-L.; Yang, M.; Lin, L.; Wu, X.-Y.; Lu, C.-Z. A unique tetranuclear Ag(I) complex emitting efficient thermally activated delayed fluorescence with a remarkably short decay time. *Dalton Trans.* **2018**, *47*, 5956–5960. [CrossRef]
25. Shafikov, M.Z.; Suleymanova, A.F.; Czerwieniec, R.; Yersin, H. Thermally activated delayed fluorescence from Ag(I) complexes: A route to 100% quantum yield at unprecedentedly short decay time. *Inorg. Chem.* **2017**, *56*, 13274–13285. [CrossRef]
26. Shafikov, M.Z.; Suleymanova, A.F.; Schinabeck, A.; Yersin, H. Dinuclear Ag(I) Complex designed for highly efficient thermally activated delayed fluorescence. *J. Phys. Chem. Lett.* **2018**, *9*, 702–709. [CrossRef]
27. Shafikov, M.Z.; Suleymanova, A.F.; Czerwieniec, R.; Yersin, H. Design strategy for Ag(I)-Based thermally activated delayed fluorescence reaching an efficiency breakthrough. *Chem. Mater.* **2017**, *29*, 1708–1715. [CrossRef]
28. Czerwieniec, R.; Leitl, M.J.; Homeier, H.H.H.; Yersin, H. Cu(I) complexes—Thermally activated delayed fluorescence. Photophysical approach and material design. *Coord. Chem. Rev.* **2016**, *325*, 2–28. [CrossRef]

29. Moussa, M.S.; Khalil, A.M.; Evariste, S.; Wong, H.-L.; Delmas, V.; Le Guennic, B.; Calvez, G.; Costuas, K.; Yam, V.W.-W.; Lescop, C. Intramolecular rearrangements guided by adaptive coordination-driven reactions toward highly luminescent polynuclear Cu(I) assemblies. *Inorg. Chem. Front.* **2020**, *7*, 1334–1344. [CrossRef]
30. Wei, Z.; Wu, X.-H.; Luo, P.; Wang, J.-Y.; Li, K.; Zang, S.-Q. Matrix coordination induced emission in a three-dimensional silver cluster-assembled material. *Chem. Eur. J.* **2019**, *25*, 2750–2756. [CrossRef]
31. Du, L.-Y.; Shi, W.-J.; Hou, L.; Wang, Y.-Y.; Shi, Q.-Z.; Zhu, Z. Solvent or temperature induced diverse coordination polymers of silver(I) sulfate and bipyrazole systems: Syntheses, crystal structures, luminescence, and sorption properties. *Inorg. Chem.* **2013**, *52*, 14018–14027. [CrossRef]
32. Durini, S.; Ardizzoia, G.A.; Therrien, B.; Brenna, S. Tuning the fluorescence emission in mononuclear heteroleptic trigonal silver(I) complexes. *New J. Chem.* **2017**, *41*, 3006–3014. [CrossRef]
33. Shafikov, M.Z.; Czerwieniec, R.; Yersin, H. Ag(I) complex design affording intense phosphorescence with a landmark lifetime of over 100 milliseconds. *Dalton Trans.* **2019**, *48*, 2802–2806. [CrossRef] [PubMed]
34. Xu, Q.-Q.; Dong, X.-Y.; Huang, R.-W.; Li, B.; Zang, S.-Q.; Mak, T.C.W. A thermochromic silver nanocluster exhibiting dual emission character. *Nanoscale* **2015**, *7*, 1650–1654. [CrossRef]
35. Hsu, C.-W.; Lin, C.-C.; Chung, M.-W.; Chi, Y.; Lee, G.-H.; Chou, P.-T.; Chang, C.-H.; Chen, P.-Y. Systematic investigation of the metal-structure–photophysics relationship of emissive d^{10}-complexes of group 11 elements: The prospect of application in organic light emitting devices. *J. Am. Chem. Soc.* **2011**, *133*, 12085–12099. [CrossRef] [PubMed]
36. Yersin, H.; Leitl, M.J.; Czerwieniec, R. TADF for singlet harvesting: Next generation OLED materials based on brightly green and blue emitting Cu(I) and Ag(I) compounds. *Proc. SPIE* **2014**, *9183*, 91830N.
37. Osawa, M.; Kawata, I.; Ishii, R.; Igawa, S.; Hashimoto, M.; Hoshino, M. Application of neutral d^{10} coinage metal complexes with an anionic bidentate ligand in delayed fluorescence-type organic light-emitting diodes. *J. Mater. Chem. C* **2013**, *1*, 4375–4383. [CrossRef]
38. Artem'ev, A.V.; Shafikov, M.Z.; Schinabeck, A.; Antonova, O.V.; Berezin, A.S.; Bagryanskaya, I.Y.; Plusnin, P.E.; Yersin, H. Sky-blue thermally activated delayed fluorescence (TADF) based on Ag(I) complexes: Strong solvation-induced emission enhancement. *Inorg. Chem. Front.* **2019**, *6*, 3168–3176. [CrossRef]
39. Osawa, M.; Hashimoto, M.; Kawata, I.; Hoshino, M. Photoluminescence properties of TADF-emitting three-coordinate silver(I) halide complexes with diphosphine ligands: A comparison study with copper(I) complexes. *Dalton Trans.* **2017**, *46*, 12446–12455. [CrossRef]
40. Titov, A.A.; Filippov, O.A.; Smol'yakov, A.F.; Godovikov, I.A.; Shakirova, J.R.; Tunik, S.P.; Podkorytov, I.S.; Shubina, E.S. Luminescent complexes of the trinuclear silver(I) and copper(I) pyrazolates supported with bis(diphenylphosphino) methane. *Inorg. Chem.* **2019**, *58*, 8645–8656. [CrossRef] [PubMed]
41. Hamze, R.; Shi, S.; Kapper, S.C.; Ravinson, D.S.M.; Estergreen, L.; Jung, M.-C.; Tadle, A.C.; Haiges, R.; Djurovich, P.I.; Peltier, J.L.; et al. "Quick-silver" from a systematic study of highly luminescent, two-coordinate, d^{10} coinage metal complexes. *J. Am. Chem. Soc.* **2019**, *141*, 8616–8626. [CrossRef]
42. Chotard, F.; Sivchik, V.; Linnolahti, M.; Bochmann, M.; Romanov, A.S. Mono-versus bicyclic carbene metal amide photoemitters: Which design leads to the best performance? *Chem. Mater.* **2020**, *32*, 6114–6122. [CrossRef]
43. Ruan, Z.-W.; Zhang, X.; Pang, A.-Y.; Dai, F.-R.; Chen, Z.-N. Blue luminescent silver(I) complexes constructed by 2-diphenylphosphinopyridine and dicyanamide or tricyanomethanide. *Inorg. Chem. Commun.* **2020**, *116*, 107916. [CrossRef]
44. Kakizoe, D.; Nishikawa, M.; Degawa, T.; Tsubomura, T. Intense blue emission and a reversible hypsochromic shift of luminescence caused by grinding based on silver(I) complexes. *Inorg. Chem. Front.* **2016**, *3*, 1381–1387. [CrossRef]
45. Yersin, H.; Czerwieniec, R.; Shafikov, M.Z.; Suleymanova, A.F. TADF material design: Photophysical background and case studies focusing on Cu^{I} and Ag^{I} complexes. *ChemPhysChem* **2017**, *18*, 3508–3535. [CrossRef] [PubMed]
46. Chakkaradhari, G.; Eskelinen, T.; Degbe, C.; Belyaev, A.; Melnikov, A.S.; Grachova, E.V.; Tunik, S.P.; Hirva, P.; Koshevoy, I.O. Oligophosphine-thiocyanate copper(I) and silver(I) complexes and their borane derivatives showing delayed fluorescence. *Inorg. Chem.* **2019**, *58*, 3646–3660. [CrossRef] [PubMed]
47. Crespo, O.; Gimeno, M.; Laguna, A.; Larraz, C. Luminescent silver(I) and copper(I) systems containing pyridyl phosphine bridges. *Z. Naturforsch. B* **2009**, *64*, 1525–1534. [CrossRef]

48. Li, S.; Han, M.; Wu, B.; Wang, J.; Zhang, F.-Q.; Zhang, X.-M. Observation of contrary thermo-responsive trend for single crystal and powder samples in mechano-, thermo- and solvato-responsive luminescent cubane $[Ag_4I_4L_4]$ cluster. *Sci. Rep.* **2017**, *7*, 13058. [CrossRef]
49. Dosen, M.; Kawada, Y.; Shibata, S.; Tsuge, K.; Sasaki, Y.; Kobayashi, A.; Kato, M.; Ishizaka, S.; Kitamura, N. Control of emissive excited states of silver(I) halogenido coordination polymers by a solid solution approach. *Inorg. Chem.* **2019**, *58*, 8419–8431. [CrossRef]
50. Cui, H.-H.; Wu, N.-N.; Wang, J.-Y.; Hu, M.-Q.; Wen, H.-M.; Chen, C.-N. Pyridyl- and pyrimidyl-phosphine-substituted [FeFe]-hydrogenase mimics: Synthesis, characterization and properties. *J. Organomet. Chem.* **2014**, *767*, 46–53. [CrossRef]
51. Miller, P.; Nieuwenhuyzen, M.; Charmant, J.P.H.; James, S.L. ROP relationships between coordination polymers and discrete complexes: Discrete bowl-shaped isomers of a 2-dimensional $\{M_4L_3\}_n$ polymer. *CrystEngComm* **2004**, *6*, 408–412. [CrossRef]
52. Cingolani, A.; Effendy; Martini, D.; Pettinari, C.; Skelton, B.W.; White, A.H. Synthesis, spectroscopic and structural characterization of novel adducts of some silver(I) salts with the ambidentate donor PPh_2py. *Inorg. Chim. Acta* **2006**, *359*, 2183–2193. [CrossRef]
53. Kuang, S.-M.; Zhang, L.-M.; Zhang, Z.-Z.; Wu, B.-M.; Mak, T.C.W. Synthesis and structural characterization of binuclear silver(I) complexes bridged by three polydentate phosphine ligands. *Inorg. Chim. Acta* **1999**, *284*, 278–283. [CrossRef]
54. Bondi, A. van der Waals volumes and radii. *J. Phys. Chem.* **1964**, *68*, 441–451. [CrossRef]
55. Zheng, J.; Yu, Y.-D.; Liu, F.-F.; Liu, B.-Y.; Wei, G.; Huang, X.-C. Modulation of argentophilic interactions by bridging amine ligands: Photoluminescence tuneable by excitation energy or temperature. *Chem. Commun.* **2014**, *50*, 9000–9002. [CrossRef] [PubMed]
56. Shekhovtsov, N.A.; Vinogradova, K.A.; Berezin, A.S.; Sukhikh, T.S.; Krivopalov, V.P.; Nikolaenkova, E.B.; Bushuev, M.B. Excitation wavelength dependent emission of silver(I) complexes with a pyrimidine ligand. *Inorg. Chem. Front.* **2020**, *7*, 2212–2223. [CrossRef]
57. Rogovoy, M.I.; Frolova, T.S.; Samsonenko, D.G.; Berezin, A.S.; Bagryanskaya, I.Y.; Nedolya, N.A.; Tarasova, O.A.; Fedin, V.P.; Artem'ev, A.V. 0D to 3D coordination assemblies engineered on silver(I) salts and 2-(alkylsulfanyl)azine ligands: Crystal structures, dual luminescence, and cytotoxic activity. *Eur. J. Inorg. Chem.* **2020**, *2020*, 1635–1644. [CrossRef]
58. Yersin, H. (Ed.) *Highly Efficient OLEDs Materials Based on Thermally Activated Delayed Fluorescence*; Wiley-VCH: Weinheim, Germany, 2019.
59. Stephens, P.J.; Devlin, F.J.; Chabalowski, C.F.; Frisch, M.J. Relaxation dynamics in the B(1/2) and C(3/2) charge transfer states of XeF in solid Ar. *J. Chem. Phys.* **1994**, *98*, 11623–11627. [CrossRef]
60. Weigend, F. Accurate Coulomb-fitting basis sets for H to Rn. *Phys. Chem. Chem. Phys.* **2006**, *8*, 1057–1065. [CrossRef]
61. Frisch, M.J.; Trucks, G.W.; Schlegel, H.B.; Scuseria, G.; Robb, M.; Cheeseman, J.; Scalmani, G.; Barone, V.; Mennucci, B.; Petersson, G.; et al. *Gaussian 09, Revision C.01*; Gaussian Inc.: Wallingford, CT, USA, 2010.
62. *CrysAlisPro 1.171.38.46*; Rigaku Oxford Diffraction: Tokyo, Japan, 2015.
63. Sheldrick, G.M. SHELXT–Integrated space-group and crystal-structure determination. *Acta Cryst. A* **2015**, *71*, 3–8. [CrossRef]
64. Sheldrick, G.M. Crystal structure refinement with SHELXL. *Acta Cryst. C* **2015**, *71*, 3–8. [CrossRef] [PubMed]

inorganics

MDPI

Article

Effect of Coordinating Solvents on the Structure of Cu(II)-4,4′-bipyridine Coordination Polymers

Marzio Rancan [1,*], Alice Carlotto [2], Gregorio Bottaro [1] and Lidia Armelao [1,2,*]

1 Institute of Condensed Matter Chemistry and Technologies for Energy (ICMATE), National Research Council (CNR), c/o Department of Chemical Sciences, University of Padova, via Marzolo 1, 35131 Padova, Italy

2 Department of Chemical Sciences, University of Padova, via Marzolo 1, 35131 Padova, Italy

* Correspondence: marzio.rancan@cnr.it or marzio.rancan@unipd.it (M.R.); lidia.armelao@unipd.it (L.A.)

Received: 23 July 2019; Accepted: 13 August 2019; Published: 19 August 2019

Abstract: Solvent can play a crucial role in the synthesis of coordination polymers (CPs). Here, this study reports how the coordinating solvent approach (CSA) can be used as an effective tool to control the nature of the final CP. This study exploited the system Cu(II)-4,4′-bipyridine coupled to different coordinating solvents, such as DMA, DMF and DMSO. This allowed the isolation and structurally characterization of four new CPs: three 2D layered networks and one 1D chain. Moreover, it was evidenced that even adventitious water can play the role of the coordinating solvent in the final CP.

Keywords: coordination polymer; MOF; CP; dimensionality control; Cu(II)-4,4′-bipyridine; dipyridil ligand; copper

1. Introduction

Coordination polymers (CPs) and metal organic frameworks (MOFs) have attracted increasing interest over the last two decades since the features of these systems are potentially useful in several cutting-edge research areas [1–3]. Countless studies have been devoted by researchers to these compounds, however there are still controversial opinions as to whether a real design of these systems can be applied [4–6]. In fact, the predictability of the final network can be a challenge since it is a consequence of the self-assembly process that involves competing, reversible and simultaneous interactions among the metal, ligand, counterion and solvent, just to mention the main chemical actors. In this context, the energy of the metal-ligand bond plays a crucial role. Metal ions and charged ligands (for instance carboxylates) can give quite strong bonds (200–400 kJ/mol ca.) paving the route to the so-called reticular chemistry that allows a good design of the final MOF [7]. This strategy has been used to develop large families of structures where the network can be designed combining the starting building blocks and at the same time, the pores size can be controlled by simply varying the length of the ligand maintaining the same network topology and obtaining isoreticular MOFs [8,9]. On the other hand, when considering weaker interactions, as for instance metal ions and neutral ligands (60–180 kJ/mol ca.), the final coordination outcome is not easy to control, both in terms of topology and network dimensionality. In fact, as the interaction energy between metal and ligand decreases, the system is more prone to be affected by other parameters, such as the counterion and the solvent. In particular, when considering Cu(II) and the 4,4′-bipyridine ligand (bpy), many different CPs and MOFs can be obtained. The isolated structure strongly depends on the counterion leading, for instance, to systems with different dimensionalities (1D, 2D and 3D) and topologies [10–14]. In addition, the solvent can play an important role. For instance, the 2D $[Cu(bpy)_2(CF_3SO_3)_2]_n$ framework can be transformed into a hydrogen bond assisted 3D framework through a solvent (H_2O) mediated process [15] and solvent dependent routes were developed to obtain 2D or 3D $[Cu(bpy)_2(CF_3SO_3)_2]_n$

networks [16]. In this context, the authors previously showed that a coordinating solvent, such as dimethyl sulfoxide (DMSO), can be used to tune and control the dimensionality of the final CP [17]. This coordinating solvent approach (CSA) promotes dimensional variability, driving the formation of Cu–bpy architectures, such as a 3D nanoporous network ($\{[Cu_2(bpy)_4(DMSO)_3(ClO_4)](ClO_4)_3{\cdot}2DMSO\}_n$, **1**) or a 1D chiral chain ($\{[Cu(bpy)_2(DMSO)_4](ClO_4)_2\}_n$, **2**) just changing the crystallization conditions (i.e., the presence of a co-solvent or evaporation rate). In solution, DMSO and bpy molecules establish a series of dynamic equilibria to coordinate the metal center during the self-assembly process. The coordinating solvent can block a different number of coordination sites leading to different CP architectures. Recently, the authors have also demonstrated that Cu–bpy bonds and coordinating solvents can be used to reversibly self-assemble mechanically interlocked CPs, such as the coordination-driven polyrotaxane-like architectures [18], when employing a Cu–metallocycle as a platform for the self-assembly of the Cu–bpy extend architectures.

Herein, this study extended CSA to other coordinating solvents (*N,N*-dimethylformamide, DMF; *N,N*-dimethylacetamide, DMA; and mixtures of them) and their effect on the final Cu–bpy based CPs were studied. Moreover, it was also demonstrated that even water can participate as a coordinating solvent during the dynamic equilibria that lead to the final CP. All the new compounds have been isolated as single crystals and structurally characterized resulting in three new 2D and one 1D CPs.

2. Results

All the CPs (**3–6**) were synthesized by dissolving $Cu(ClO_4)_2{\cdot}6(H_2O)$ in a coordinating solvent and adding bpy in a 1:2 ratio. The solvent evaporation or diffusion of a co-solvent allowed isolating single crystals of the compounds. All the structures were solved by the single-crystal X-ray crystallographic method and their phase purity confirmed by powder X-ray diffraction (PXRD, Figure S1). The important refinement and geometric parameters are shown in Table 1.

Table 1. Crystal data and structures refinement.

Compound	3	4	5	6
Formula	$C_{28}H_{38}Cl_2CuN_6O_{12}$	$C_{46}H_{46}Cl_2CuN_{10}O_{10}$	$C_{40}H_{36}Cl_2CuN_8O_{11}$	$C_{20}H_{34}Cl_2CuN_4O_{12}S_2$
Formula weight	785.08	1033.37	939.21	721.07
Temperature/K	301(2)	301(3)	301.2(8)	299.6(6)
Crystal system	monoclinic	monoclinic	monoclinic	monoclinic
Space group	$C2/c$	$P2_1/n$	$P2/n$	$I2/a$
a/Å	14.4924(10)	10.2921(4)	12.9264(5)	17.9524(17)
b/Å	11.1385(7)	15.7544(4)	11.1702(4)	11.0913(7)
c/Å	22.2237(13)	15.3205(5)	15.0564(10)	16.3512(15)
α/°	90	90	90	90
β/°	91.562(7)	106.641(4)	106.559(6)	100.154(8)
γ/°	90	90	90	90
Volume/Å^3	3586.1(4)	2380.11(14)	2083.83(19)	3204.8(5)
Z	4	2	2	4
Goodness-of-fit on F^2	1.117	1.024	1.081	1.035
Final R indexes $[I >= 2\sigma(I)]$	$R_1 = 0.0575$, $wR_2 = 0.1683$	$R_1 = 0.0338$ $wR_2 = 0.0914$	$R_1 = 0.0473$ $wR_2 = 0.1279$	$R_1 = 0.0340$ $wR_2 = 0.0843$
Largest diff. peak/hole/e Å^{-3}	0.60/−0.40	0.35/−0.25	0.62/−0.71	0.35/−0.32
CCDC	1942113	1942114	1942115	1942116

By using DMA, light blue single crystals of $\{[Cu(bpy)_2(H_2O)_2](ClO_4)_2{\cdot}2DMA\}_n$ (**3**) were obtained in good yield (70% ca.). An X-ray analysis evidenced that four bpy molecules in the equatorial plane and two water molecules in apical positions coordinated the copper center, Figure 1a. The copper atom has a Jahn–Teller distorted octahedral coordination with Cu–N distances of 2.027(4), 2.036(6) Å and 2.047(6) (for Cu1–N1, Cu1–N2 and Cu1–N3, respectively) and a Cu1–O1 bond length of 2.482(2) Å. The water molecule forms H-bonds with a ClO_4^- anion (2.275(9) Å) and with a DMA molecule (1.972(5) Å). The network develops as a 2D CP, leading to layers of grids composed of repeating squares with

Cu-bpy–Cu sides (Figure 1b). Two slightly different Cu-bpy–Cu distances can be found, one of 11.1385(7) Å coinciding with the *b* axis and the other of 11.1165(7) Å along the *c* axis equal to half its length. The 2D layers are at a distance of *a*/2 (7.3462(5) Å) and mismatched along the *b* axis with a value of *b*/2 (Figure 1c). This packing of the grids layers leads to the formation of two different alternating channels (Figure 1d) that occupy the 66% of unit cell volume. In one case, the plane containing the bpy molecules of a grid is parallel to the grow direction of the channel. In the other one, that plane is perpendicular to the channel direction. Perchlorate anions and DMA molecules that lay between the grid layers fill the channels.

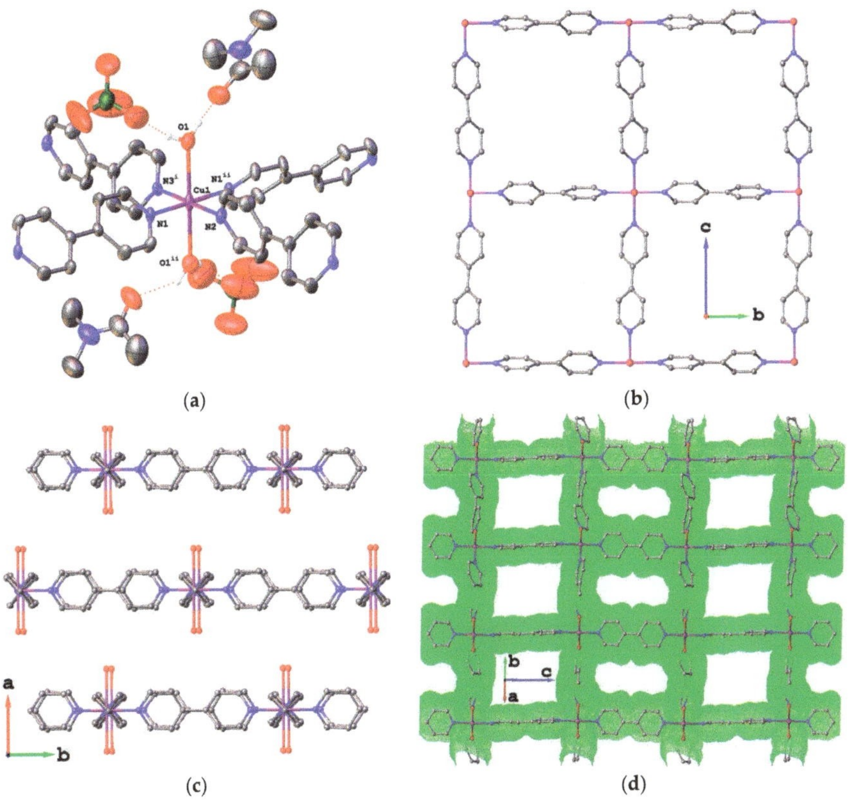

Figure 1. (**a**) Coordination environment of the copper center in compound **3** (thermal ellipsoids drawn at the 50% probability level). (**b**) 2D network (single layer); (**c**) view along the *ab* plane of alternating and mismatched grids; (**d**) channels formed by the 2D layers (green points). Color code: Cu, purple; O, red; N, blue; C, grey; Cl, green; H, white. Anions, DMA molecules and H atoms omitted for clarity in the packing figures. Symmetry operations: $i = 1 - x, -1 + y, 1/2 - z$; $ii = 1 - x, +y, 1/2 - z$.

When DMF was used in place of DMA, the solvent evaporation at open air led to a major crop (yield 40% ca.) of deep blue single crystals of $\{[Cu(bpy)_2(DMF)_2](ClO_4)_2{\cdot}2(bpy)\}_n$ (**4**) along with a minor fraction (yield 3% ca.) of light blue single crystals of $\{[Cu(bpy)_2(H_2O)_2](ClO_4)_2{\cdot}2(bpy){\cdot}2(H_2O)\}_n$ (**5**).

Considering compound **4**, four bpy molecules in the equatorial plane coordinate the copper center and the apical positions are occupied by DMF molecules (Figure 2a). Copper has a Jahn–Teller distorted octahedral coordination with Cu–N distances of 2.0250(12), 2.0564(12) Å (for Cu1–N1 and Cu1–N2, respectively) and a Cu1–O1 bond length of 2.5180(12) Å. Even **4** is a 2D CP with layers of grids (Figure 2b) composed of repeating squares with Cu–bpy–Cu sides of 11.1640(3) Å and one

Cu···Cu diagonal coinciding with *b* axis (15.7544(4) Å). The 2D layers in **4** are at a distance of 10.3 Å and mismatched of 5.6 Å (Figure 2c). This layer packing leads to the formation of only one kind of channels (Figure 2d) that occupies 57% of the unit cell volume. These channels are filled by the ClO_4^- anions that lay between the grid layers and by uncoordinated bpy molecules hosted in the Cu–bpy squares. In particular, each square hosts two bpy molecules that interacts with the CP network with a series of CH···π interactions and among them by π···π stacking (Figure 2e,f).

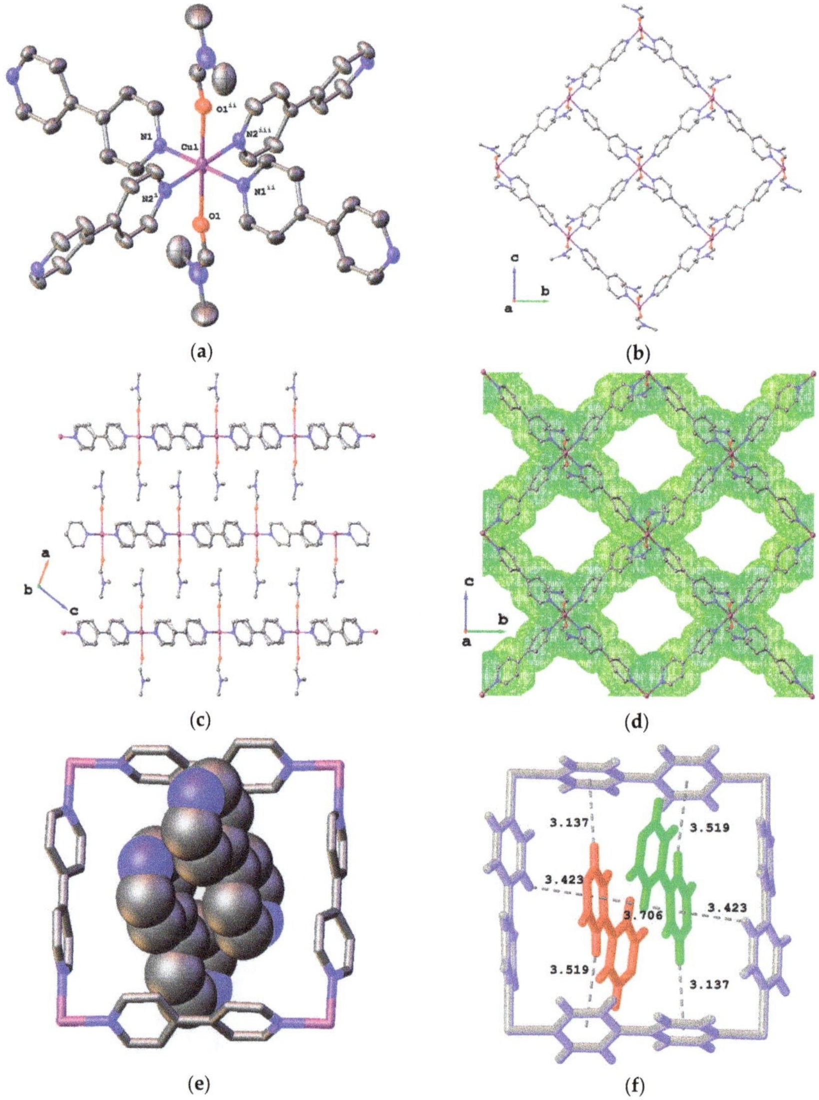

Figure 2. (**a**) Coordination environment of the copper center in compound **4** (thermal ellipsoids drawn at the 50% probability level). (**b**) 2D network (single layer); (**c**) view along the *ac* plane of alternating and mismatched grids; (**d**) channels formed by the 2D layers (green points). (**e**) Two uncoordinated bpy molecules hosted by a Cu–bpy square. (**f**) CH···π interactions and π···π stacking in the host-guest ensemble. Color code: Cu, purple; O, red; N, blue; C, grey. Anions and H atoms omitted for clarity. Symmetry operations: $i = 2 - x, 1 - y, 1 - z$; $ii = 1/2 + x, 3/2 - y, 1/2 + z$; $iii = 3/2 - x, 1/2 + y, 1/2 - z$.

In CP **5**, the coordination environment of the Cu^{2+} ions is similar, as in CP **3**. The four bpy molecules in the equatorial plane and the two water molecules in apical positions coordinate the copper center (Figure 3a). The metal atom has a Jahn–Teller distorted octahedral coordination with Cu–N distances of 2.0670(19), 2.041(3) Å and 2.045(3) (for Cu1–N1, Cu1–N2 and Cu1–N3, respectively) and a Cu1–O1 bond length of 2.4965(19) Å. Similar to **3** and **4**, the network develops as a 2D CP, leading to layers of grids composed of repeating squares with Cu–bpy–Cu sides (Figure 3b). Further in CP **5**, the apical water molecules form H-bonds with a ClO_4^- anion (2.128(3) Å) and with an uncoordinated bpy molecule (2.005(3) Å). The side of the bpy molecule not involved in the H-bond is diagonally pointing towards the center of a Cu–bpy square of a second grid layer. In this case, the starting DMF coordinating solvent is not found in the structure. Two slightly different Cu–bpy–Cu distances can be found: one of 11.1702(4) Å coinciding with the *b* axis and the other of 11.2841(6) Å. The 2D layers in **5** are at a closer distance (8.3 Å) compared to **3** and only slightly mismatched (Figure 3c). This layer packing leads to the formation of only one kind of channels (Figure 3d) that occupies 70% of the unit cell volume. Perchlorate anions and bpy molecules, that lay between and inside (bpy) the grid layers, fill those channels.

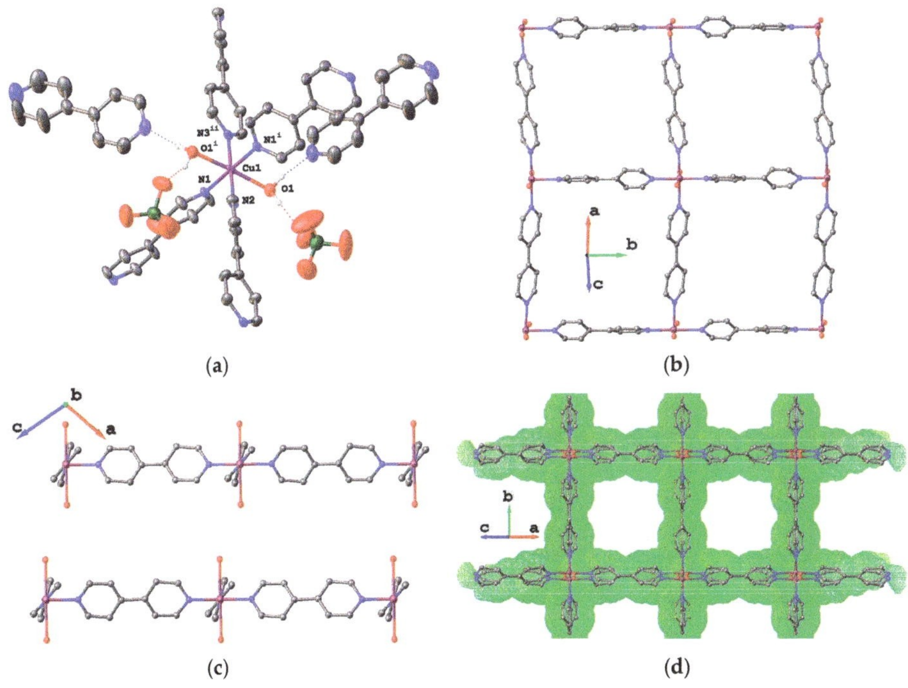

Figure 3. (**a**) Coordination environment of the copper center in compound **5** (thermal ellipsoids drawn at the 50% probability level). (**b**) 2D network (single layer); (**c**) view along the *ac* plane of alternating and slightly mismatched grids; (**d**) channels formed by the 2D layers (green points). Color code: Cu, purple; O, red; N, blue; C, grey; Cl, green; H, white. Anions, uncoordinated bpy molecules and H atoms omitted for clarity in the packing figures. Symmetry operations: $i = 1/2 - x, +y, 1/2 - z$; $ii = +x, -1 - y, +z$.

It is worthy to note that both CPs **3** and **5** are obtained with coordinated water molecules instead of the solvents, DMA or DMF. This is an important difference compared to what happens when using DMSO as a coordinating solvent [17]. In that case, slow evaporation led to a Cu–bpy based 3D CP (**1**) while faster evaporation gave a 1D chiral chain (**2**). In both cases, the only solvent coordinating the

Cu center was DMSO. However, it was postulated that in the CSA, the role of water as competitive coordinating solvent may not be excluded. As a matter of fact, synthesis with DMA and DMF confirmed our hypothesis. The sources of water can be found in the hydration water molecules of the starting $Cu(ClO_4)_2$, in the non-anhydrous solvents, or even in the evaporation performed at open air.

Finally, the employment of mixtures of coordinating solvents was explored. DMA/DMF, DMF/DMSO and DMA/DMSO 1:1 solutions evaporation did not give any crystalline product suitable for single crystal X-ray diffraction. Very slow vapor diffusion of diethylether in a DMF/DMSO mixture led to light blue single crystals of $\{[Cu(bpy)_2(DMF)(DMSO)](ClO_4)_2\}_n$ (**6**, Figure 4) in low yields (10% ca.). The X-ray structure shows that copper has a Jahn–Teller distorted octahedral coordination with the equatorial positions taken by two bpy molecules (Cu1–N1 and Cu1–N2 2.004(2) Å) and two DMSO molecules (Cu1–O1 2.0114(15) Å). The axial positions are occupied by two DMF molecules (Cu1–O2 2.3662(17) Å). The CP develops as a 1D chain with a Cu–bpy–Cu distance of 11.0913(7) Å coinciding with *b* axis. The chain grow linearly since the two bpy molecules lay in equatorial *trans* positions. On the contrary, when using only DMSO, the bpy ligands take two equatorial *cis* positions leading to a helicoidal 1D CP isolated as enantiopure single crystals (**2**) [17].

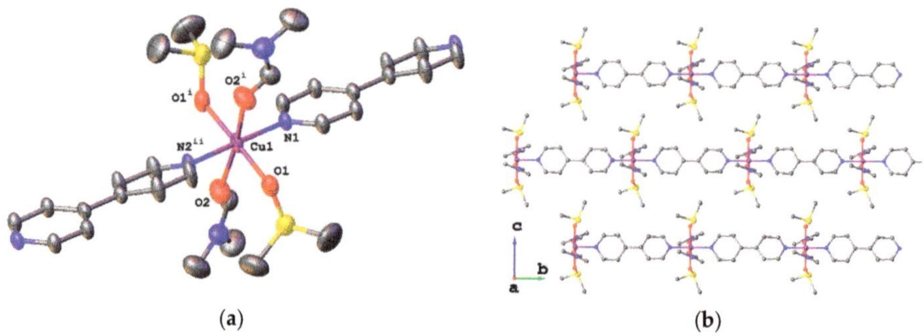

Figure 4. (**a**) Coordination environment of the copper center in compound **6** (thermal ellipsoids drawn at the 50% probability level). (**b**) 1D chains along the plane *cb*. Color code: Cu, purple; O, red; N, blue; C, grey; S, yellow. Anions and H atoms omitted for clarity. Symmetry operations: $i = 3/2 - x, +y, 1 - z$; $ii = 3/2 - x, -1 + y, 1 - z$.

3. Discussion

By employing strong coordinating solvents coupled to a neutral ligand, such as bpy, competitive coordination equilibria are introduced in the self-assembly process. The core concept of the CSA is depicted in Figure 5. The results obtained in the authors current and former [17] studies on the effect of coordinating solvents towards the final Cu–bpy based CPs are summarized in Table 2.

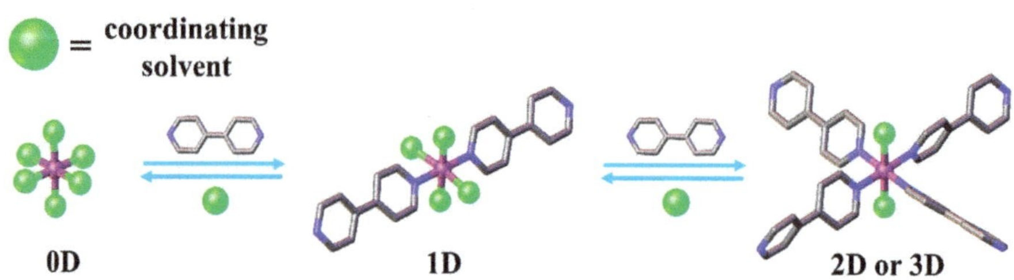

Figure 5. Scheme of the competing equilibria between the coordinating solvent and bridging ligand leading to different coordination polymers (CPs) through the coordinating solvent approach (CSA).

Table 2. The effect of coordinating solvents towards the final Cu–bpy based CPs.

Solvent.	CP	Coordinated Solvent	CP Dimensionality
DMSO [1]	**1** [17]	2 DMSO	3D
DMSO [2]	**2** [17]	4 DMSO	1D
DMA [1]	**3**	2 H_2O	2D
DMF [1]	**4** [3]	2 DMF	2D
DMF [1]	**5** [4]	2 H_2O	2D
DMSO/DMF [2]	**6**	2 DMSO + 2 DMF	1D

[1] slow evaporation; [2] in the presence of a non-coordinating solvent; [3] major product; [4] minor product.

The competitive equilibria are driven towards the CP by solvent evaporation or by introducing a co-solvent to change the medium features and to induce crystallization. Hence, when dissolving the metal center in a coordinating solvent, the process leading to the formation of the final CP can be described as the subsequent substitution of the coordinated solvent molecules by the bridging ligands. Thus, the solvent becomes itself a ligand that can compete with the bpy. This allows obtaining CPs with different networks and dimensionalities according to the solvent nature, the crystallization technique and to the remaining number of the coordinated solvent molecules versus the divergent ligands. Competitive species are present and this can lead to obtaining byproducts in low yields as in the case of CP **5**. This byproduct of CP **4** appears in the last stages of solvent evaporation and can be easily avoided by stopping the evaporation before completeness. Some general trends for the CSA applied to Cu–bpy systems can be found. The bridging bpy molecules always occupy the equatorial positions of the Cu octahedral coordination sphere. When four solvent molecules coordinate the copper (II) ion, 1D chains can be obtained, with the bpy molecules either in *trans* (CP **6**) or *cis* (CP **2**) positions. The presence of two apical solvent molecules in the coordination sphere always leads to the same structural motif, i.e., Cu–bpy squares forming extended grid-like architectures that can develop towards 2D (CPs **3**, **4**, **5**) or 3D (CP **1**) networks. Finally, it is worth noting that, when using DMA or DMF, it is likely to obtain CPs with coordinated water molecules instead of DMA or DMF ones, either as a single isolated product (**3**) or as an impurity (**5**).

4. Materials and Methods

4.1. Synthesis

The reagents were purchased from Sigma-Aldrich (St. Louis, MO, USA) and used as received. The elemental analyses were carried out with a Flash 2000 Thermo Scientific analyzer (Thermo Fisher Scientific, Cambridge, UK) at the Department of Chemical Sciences of the University of Padova.

4.1.1. Synthesis of $\{[Cu(bpy)_2(H_2O)_2](ClO_4)_2{\cdot}2DMA\}_n$ (3)

$Cu(ClO_4)_2{\cdot}6H_2O$ (37 mg, 0.1 mmol) was dissolved in 10 mL of DMA in a large beaker and 4,4′-bipyridine (bpy, 30 mg, 0.2 mmol) was added. The slow evaporation of the solvent led to light blue single crystals suitable for an X-ray analysis. Yield 55 mg, 70% (based on copper). The elemental analysis for $C_{28}H_{38}Cl_2CuN_6O_{12}$, exp (%): C 42.53, N 10.58, H 4.92; calc (%): C 42.84, N 10.70, H 4.88.

4.1.2. Synthesis of $\{[Cu(bpy)_2(DMF)_2](ClO_4)_2{\cdot}2(bpy)\}_n$ (4) and $\{[Cu(bpy)_2(H_2O)_2](ClO_4)_2{\cdot}2(bpy){\cdot}2(H_2O)_2\}_n$ (5)

$Cu(ClO_4)_2{\cdot}6H_2O$ (37 mg, 0.1 mmol) was dissolved in 10 mL of DMF in a large beaker and 4,4′-bipyridine (bpy, 30 mg, 0.2 mmol) was added. The slow evaporation of the solvent led to deep blue single crystals of **4**. If the solvent was left to completely evaporate, a second kind of light blue single crystals appeared in the last evaporation stages as an impurity (**5**, yield 3 mg, 3% ca.). Due to their different color, the two species can be easily manually separated. To avoid crystallization of CP **5**, single crystals of compound **4** was removed from the solution before complete evaporation with a yield of 43 mg, 40% (based on copper). The elemental analysis for **4**, $C_{46}H_{46}Cl_2CuN_{10}O_{10}$, exp (%): C 53.31, N 13.44, H 4.56; calc (%): C 53.47, N 13.55, H 4.49. The elemental analysis for **5**, $C_{40}H_{40}Cl_2CuN_8O_{11}$, exp (%): C 51.72, N 11.98, H 4.21; calc (%): C 51.93, N 11.88, H 4.27.

4.1.3. Synthesis of {[Cu(bpy)$_2$(DMF)(DMSO)](ClO_4)$_2$}$_n$ (**6**)

$Cu(ClO_4)_2{\cdot}6H_2O$ (37 mg, 0.1 mmol)) was dissolved in 1 mL of a DMSO/DMF (1:1) solution and 4,4′-bipyridine (bpy, 30 mg, 0.2 mmol) was added. The very slow diffusion of diethylether vapor gave light blue single crystals after 6 weeks. Yield 8 mg, 10% (based on copper). The elemental analysis for $C_{20}H_{34}Cl_2CuN_4O_{12}S_2$, exp (%): C 33.37, N 8.01, S 9.05, H 4.83; calc (%): C 33.31, N 7.77, S 8.89, H 4.75.

4.2. Crystal Structure Determination

The data were collected using an Oxford Diffraction Gemini E diffractometer (Oxford Diffraction, Oxfordshire, England), equipped with a 2K × 2K EOS CCD area detector and sealed-tube Enhance (Mo) and (Cu) X-ray sources. The single crystals of compounds were fastened on the top of a Lindemann glass capillary. The data were collected by means of the ω-scans technique using graphite-monochromated radiation. The detector distance was set at 45 mm. The diffraction intensities were corrected for Lorentz/polarization effects as well as with respect to the absorption. The empirical multi-scan absorption corrections using equivalent reflections were performed with the scaling algorithm SCALE3 ABSPACK. The data reduction, finalization and cell refinement were carried out through the CrysAlisPro software (1.171.38.46, Rigaku Oxford Diffraction, Rigaku Corporation, Oxford, UK). The accurate unit cell parameters were obtained by least squares refinement of the angular settings of the strongest reflections, chosen from the whole experiment. The structures were solved with *Olex2* [19] by using *ShelXT* [20] structure solution program by Intrinsic Phasing and refined with the *ShelXL* [21] refinement package using least-squares minimization. In the last cycles of refinement, non-hydrogen atoms were refined anisotropically. Hydrogen atoms were included in the calculated positions, and a riding model was used for their refinement. For CP **3**, the indexing of the collected data and frame inspections clearly showed twining signals. The data were processed with the twin/multicristal routine of the CrysAlisPro software. The logout of the twin data reduction is given as Supplementary Materials. The twin data reduction with two components allowed solving the structure. The two components were rotated at 180° around the [001] vector. The final refined BASF parameter was 0.389(2). The specific refinement details for each compounds are embedded in their CIF files given as Supplementary Materials and that have been deposited with the Cambridge Crystallographic Data Centre as supplementary publication (CCDC 1942113–1942116). Copies of the data can be obtained free of charge on application to the CCDC, 12 Union Road, Cambridge CB2 1EZ, U.K. (fax, (+44) 1223 336033; e-mail, deposit@ccdc.cam.ac.uk).

4.3. Powder X-ray Diffraction (PXRD)

The PXRD patterns of CP **3** and **4** were collected with a Bruker D8 Advance diffractometer (Bruker AXS, Karlsruhe, Germany), in Bragg–Brentano geometry, using Cu Kα. The patterns were acquired in the 5°–50° 2θ range (0.03°/step and 10 s/step). The PXRD patterns of CP **5** and **6** were collected with the powder diffraction tool of an Oxford Diffraction Gemini E diffractometer using Cu Kα. The powder diffraction images (20 frames) were collected over 100 s exposition time with a 90 degrees ϕ rotation and a detector distance of 120 mm.

5. Conclusions

This study analyzed and studied the effect of coordinating solvents on the final outcome of the self-assembly process towards Cu(II)-4,4′-bipyridine based coordination polymers. Solvents such as DMA, DMF and DMSO can compete with the bridging ligand in occupying two or four coordination sites of the metal center. In this competition, also adventitious water can participate. This allowed the access to different coordination polymers. In particular, it was found that the presence of four coordinating solvent molecules lead to 1D polymers, while two solvent molecules, in the apical sites, always gave the same fundamental unit (Cu–bpy square) that developed in grid-like structures to give 2D or 3D networks. Our results show that the coordinating solvent approach (CSA) can be used as an effective tool to modulate and control the dimensionality, composition and network of coordination polymers.

Supplementary Materials: The following are available online at http://www.mdpi.com/2304-6740/7/8/103/s1, Figure S1: PXRD patterns; CIF and checkCIF files of the crystal structures, and twin logout for CP **3**.

Author Contributions: Conceptualization, M.R.; validation M.R. and L.A.; investigation M.R., A.C. and G.B.; writing—original draft preparation, M.R. and L.A.; writing—review and editing, all authors; funding acquisition, M.R. and L.A.

Funding: This research was funded by the University of Padova (grant: P-DISC #CARL-SID17 BIRD2017-UNIPD), project CHIRoN and by Ministero Istruzione Università e Ricerca, MIUR (PRIN 2015, 20154X9ATP, Progetti di Ricerca di Interesse Nazionale. APC was sponsored by MDPI.

Conflicts of Interest: The authors declare no conflict of interest. The funders had no role in the design of the study; in the collection, analyses, or interpretation of data; in the writing of the manuscript, or in the decision to publish the results.

References

1. Kitagawa, S.; Kitaura, R.; Noro, S. Functional Porous Coordination Polymers. *Angew. Chem. Int. Ed.* **2004**, *43*, 2334–2375. [CrossRef] [PubMed]
2. Furukawa, H.; Cordova, K.E.; O'Keeffe, M.; Yaghi, O.M. The Chemistry and Applications of Metal–Organic Frameworks. *Science* **2013**, *341*, 1230444. [CrossRef] [PubMed]
3. Zhou, H.-C.; Long, J.R.; Yaghi, O.M. Introduction to Metal–Organic Frameworks. *Chem. Rev.* **2012**, *112*, 673–674. [CrossRef] [PubMed]
4. Jansen, M.; Schön, J.C. "Design" in Chemical Synthesis—An Illusion? *Angew. Chem. Int. Ed.* **2006**, *45*, 3406–3412. [CrossRef] [PubMed]
5. O'Keeffe, M. Design of MOFs and intellectual content in reticular chemistry: A personal view. *Chem. Soc. Rev.* **2009**, *38*, 1215–1217. [CrossRef]
6. Goesten, M.G.; Kapteijn, F.; Gascon, J. Fascinating chemistry or frustrating unpredictability: Observations in crystal engineering of metal–organic frameworks. *CrystEngComm* **2013**, *15*, 9249–9257. [CrossRef]
7. Yaghi, O.M. Reticular Chemistry—Construction, Properties, and Precision Reactions of Frameworks. *J. Am. Chem. Soc.* **2016**, *138*, 15507–15509. [CrossRef]
8. Eddaoudi, M.; Kim, J.; Rosi, N.; Vodak, D.; Wachter, J.; O'Keeffe, M.; Yaghi, O.M. Systematic design of pore size and functionality in isoreticular MOFs and their application in methane storage. *Science* **2002**, *295*, 469–472. [CrossRef]
9. Deng, H.; Grunder, S.; Cordova, K.E.; Valente, C.; Furukawa, H.; Hmadeh, M.; Gándara, F.; Whalley, A.C.; Liu, Z.; Asahina, S.; et al. Large-pore apertures in a series of metal–organic frameworks. *Science* **2012**, *336*, 1018–1023. [CrossRef]
10. Noro, S.; Kitaura, R.; Kondo, M.; Kitagawa, S.; Ishii, T.; Matsuzaka, H.; Yamashita, M. Framework Engineering by Anions and Porous Functionalities of Cu(II)/4,4′-bpy Coordination Polymers. *J. Am. Chem. Soc.* **2002**, *124*, 2568–2583. [CrossRef]
11. Carlucci, L.; Cozzi, N.; Ciani, G.; Moret, M.; Proserpio, D.M.; Rizzato, S. A three-dimensional nanoporous flexible network of 'square-planar' copper(II) centres with an unusual topology. *Chem. Commun.* **2002**, 1354–1355. [CrossRef]
12. Blake, A.J.; Hill, S.J.; Hubberstey, P.; Li, W.-S. Rectangular grid two-dimensional sheets of copper(II) bridged by both co-ordinated and hydrogen bonded 4,4′-bipyridine (4,4′-bipy) in [Cu(μ-4,4′-bipy)(H_2O)$_2$ (FBF_3)$_2$]·4,4′-bipy. *J. Chem. Soc. Dalt. Trans.* **1997**, 913–914. [CrossRef]
13. Masciocchi, N.; Cairati, P.; Carlucci, L.; Mezza, G.; Ciani, G.; Sironi, A. Ab-initio X-ray powder diffraction structural characterization of co-ordination compounds: Polymeric [{MX_2(bipy)}$_n$] complexes (M = Ni or Cu; X = Cl or Br; bipy = 4,4′-bipyridyl). *J. Chem. Soc. Dalt. Trans.* **1996**, 2739–2746. [CrossRef]

14. Rizzato, S.; Moret, M.; Beghi, F.; Lo Presti, L. Crystallization and structural properties of a family of isotopological 3D-networks: The case of a 4,4′-bipy ligand–M^{2+} triflate system. *CrystEngComm* **2018**, *20*, 3784–3795. [CrossRef]
15. Komori-Orisaku, K.; Hoshino, K.; Yamashita, S.; Koide, Y. Water Molecules as Binders in Transformation of 2D Coordination Polymer $[Cu(4,4'\text{-bpy})_2(OTf)_2]_n$ into Parallel Aligned 3D Architectures. *Bull. Chem. Soc. Jpn.* **2010**, *83*, 276–278. [CrossRef]
16. Kondo, A.; Kajiro, H.; Noguchi, H.; Carlucci, L.; Proserpio, D.M.; Ciani, G.; Kato, K.; Takata, M.; Seki, H.; Sakamoto, M.; et al. Super Flexibility of a 2D Cu-Based Porous Coordination Framework on Gas Adsorption in Comparison with a 3D Framework of Identical Composition: Framework Dimensionality-Dependent Gas Adsorptivities. *J. Am. Chem. Soc.* **2011**, *133*, 10512–10522. [CrossRef] [PubMed]
17. Rancan, M.; Armelao, L. Exploiting dimensional variability in coordination polymers: Solvent promotes reversible conversion between 3D and chiral 1D architectures. *Chem. Commun.* **2015**, *51*, 12947–12949. [CrossRef]
18. Truccolo, G.; Tessari, Z.; Tessarolo, J.; Quici, S.; Armelao, L.; Rancan, M. A Cu(II) metallocycle for the reversible self-assembly of coordination-driven polyrotaxane-like architectures. *Dalt. Trans.* **2018**, *47*, 12079–12084. [CrossRef]
19. Dolomanov, O.V.; Bourhis, L.J.; Gildea, R.J.; Howard, J.A.K.; Puschmann, H. *OLEX2*: A complete structure solution, refinement and analysis program. *J. Appl. Crystallogr.* **2009**, *42*, 339–341. [CrossRef]
20. Sheldrick, G.M. IUCr *SHELXT*—Integrated space-group and crystal-structure determination. *Acta Crystallogr. Sect. A Found. Adv.* **2015**, *71*, 3–8. [CrossRef]
21. Sheldrick, G.M. Crystal structure refinement with *SHELXL*. *Acta Crystallogr. Sect. C Struct. Chem.* **2015**, *71*, 3–8. [CrossRef]

inorganics

MDPI

Article

New Coordination Polymers of Zinc(II), Copper(II) and Cadmium(II) with 1,3-Bis(1,2,4-triazol-4-yl)adamantane

Nertil Xhaferaj [1,2], Aurel Tăbăcaru [3,*], Marco Moroni [4], Ganna A. Senchyk [5], Kostiantyn V. Domasevitch [5], Claudio Pettinari [2,6] and Simona Galli [4,7,*]

1 Department of Food Technology, Faculty of Food and Biotechnology, Agricultural University of Tirana, Kodër Kamëz, SH1, 1000 Tirana, Albania; nertil.xhaferaj@gmail.com
2 School of Pharmacy, University of Camerino, Via S. Agostino 1, 62032 Camerino, Italy; claudio.pettinari@unicam.it
3 Department of Chemistry, Physics and Environment, Faculty of Sciences and Environment, "Dunarea de Jos" University of Galati, 111 Domneasca Street, 800201 Galati, Romania
4 Dipartimento di Scienza e Alta Tecnologia, Università deli Studi dell'Insubria, Via Valleggio 11, 22100 Como, Italy; mmoroni@studenti.uninsubria.it
5 Inorganic Chemistry Department, Taras Shevchenko National University of Kyiv, 64 Volodimirska Street, 01033 Kyiv, Ukraine; senchyk.ganna@gmail.com (G.A.S.); dk@univ.kiev.ua (K.V.D.)
6 Istituto di Chimica dei Composti Organometallici (ICCOM-CNR), Via Madonna del Piano 10, 50019 Sesto Fiorentino, Italy
7 Consorzio Interuniversitario Nazionale per la Scienza e Tecnologia dei Materiali, Via Giusti 9, 50121 Firenze, Italy
* Correspondence: aurel.tabacaru@ugal.ro (A.T.); simona.galli@uninsubria.it (S.G.)

Received: 16 October 2020; Accepted: 30 October 2020; Published: 6 November 2020

check for updates

Abstract: The new coordination polymers (CPs) $[Zn(tr_2ad)Cl_2]_n$, $\{[Cu(tr_2ad)Cl]Cl{\cdot}4H_2O\}_n$, $[Cd_2(tr_2ad)Cl_4]_n$, $\{[Cu(tr_2ad)(NO_3)](NO_3)\}_n$ and $\{[Cd(tr_2ad)(NO_3)](NO_3){\cdot}H_2O\}_n$ were obtained in the form of air- and moisture-stable microcrystalline powders by the solvothermal reactions of zinc(II), copper(II) and cadmium(II) chlorides or nitrates with the ligand 1,3-bis(1,2,4-triazol-4-yl)adamantane (tr_2ad). Investigation of the thermal behaviour assessed the thermal stability of these CPs, with $[Cd_2(tr_2ad)Cl_4]_n$ starting to decompose only around 365 °C. As retrieved by powder X-ray diffraction, while $[Zn(tr_2ad)Cl_2]_n$ features 1-D chains along which the metal centre shows a tetrahedral geometry and the spacer is exo-bidentate, the other CPs contain 2-D double-layers in which the metal ions possess an octahedral stereochemistry and the linker is exo-tetradentate. A comparative structural analysis involving known coordination compounds containing the tr_2ad ligand enabled us to disclose (i) the versatility of the ligand, as far as the coordination modes are concerned; (ii) the variability in crystal structure dimensionality, ranging from 1-D to 3-D; (iii) the fact that, to the best of our knowledge, $[Zn(tr_2ad)Cl_2]_n$ is the first Zn^{II}-based CP containing the tr_2ad spacer.

Keywords: coordination polymers; poly(azolate) spacers; 1,3-bis(1,2,4-triazol-4-yl)adamantane; zinc; copper; cadmium; crystal structure

1. Introduction

Since the discovery that metal ions and organic ligands can act as connectors and spacers, respectively, to generate infinite frameworks [1], the chemistry of coordination polymers (CPs) [2–5], including the subclass of metal–organic frameworks (MOFs) [6–10], has recorded a rapid growth, due to the plethora of functional properties they were found to possess. One of the main advantages of CPs and MOFs is the possibility to modulate their chemical composition, crystal structure and functional properties through a modification of the metal ion and/or the organic spacer. In view of their potential applications, CPs and MOFs appear as interesting platforms which may offer sustainable solutions in fields of major economical, technological and environmental importance, e.g., gas storage and separation [11], catalysis [12], luminescence [13,14], conductivity [15], magnetism [16], sensing [17–19] and biomedicine [20]. The successful preparation of CPs has generally relied on organic ligands from the class of poly(carboxylic) acids [21–23], pyrazines and bipyridines [21–24], phosphonic acids [25] and poly(azoles) [26–28].

Among the nitrogen-donor ligands from the class of poly(azoles), attention has been paid also to 1,2,4-triazolyl derivatives, due to their electron-donating ability and rich coordination chemistry. As a representative example, they can provide the N^1,N^2-bridging between two adjacent metal ions [29] in the same manner as pyrazolates do [30]. Based on the coordination modes they can adopt, 1,2,4-triazolyl ligands have been exploited in building up polynuclear and polymeric coordination compounds [31–35]. This is also the case of the ditopic ligand 1,3-bis(1,2,4-triazol-4-yl)adamantane (tr_2ad, Scheme 1) which, although at present less explored, provides an attractive platform for crystal engineering.

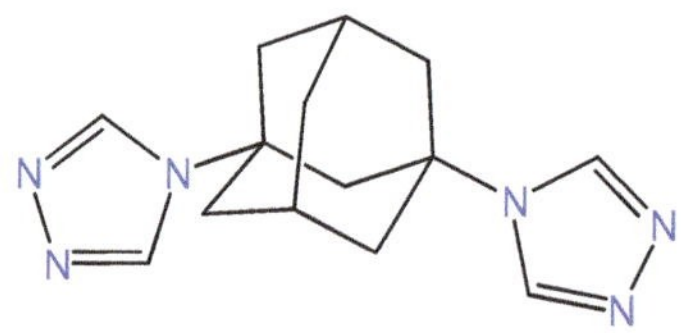

Scheme 1. Molecular structure of 1,3-bis(1,2,4-triazol-4-yl)adamantane (tr_2ad).

Aiming at enlarging and diversifying the library of tr_2ad-based coordination frameworks, we report hereafter on the synthesis, thermal behavior and structural characterization of the five new compounds $[Zn(tr_2ad)Cl_2]_n$, $\{[Cu(tr_2ad)Cl]Cl{\cdot}4H_2O\}_n$, $[Cd_2(tr_2ad)Cl_4]_n$, $\{[Cu(tr_2ad)(NO_3)](NO_3)\}_n$ and $\{[Cd(tr_2ad)(NO_3)](NO_3){\cdot}H_2O\}_n$. The crystal and molecular structures of the anhydrous and trihydrate tr_2ad ligand are also described.

2. Results and Discussion

2.1. Synthesis and Preliminary Characterization

A detailed description of the synthesis of the tr_2ad ligand, including analytical details on the intermediates never reported before, is provided in the Supporting Information.

Several screening reactions, involving the adoption of synthetic conditions differing in solvent, metal-to-ligand ratio, temperature, and/or time, were carried out in order to successfully obtain microcrystalline batches of the tr_2ad-based CPs $[Zn(tr_2ad)Cl_2]_n$, $\{[Cu(tr_2ad)Cl]Cl{\cdot}4H_2O\}_n$, $[Cd_2(tr_2ad)Cl_4]_n$, $\{[Cu(tr_2ad)(NO_3)](NO_3)\}_n$ and $\{[Cd(tr_2ad)(NO_3)](NO_3){\cdot}H_2O\}_n$. Scheme 2 shows the synthetic conditions fruitfully used for their isolation.

Compounds $[Zn(tr_2ad)Cl_2]_n$ and $\{[Cu(tr_2ad)Cl]Cl{\cdot}4H_2O\}_n$ were isolated by carrying out a solvothermal reaction among zinc(II) chloride dihydrate and anhydrous copper(II) chloride, respectively, and tr_2ad in the 2:1 molar ratio (DMF, 150 °C, 24 h). Also, the formation of compounds $[Cd_2(tr_2ad)Cl_4]_n$, $\{[Cu(tr_2ad)(NO_3)](NO_3)\}_n$ and $\{[Cd(tr_2ad)(NO_3)](NO_3){\cdot}H_2O\}_n$ required the application of solvothermal conditions, reacting anhydrous cadmium(II) chloride, copper(II) nitrate hemipentahydrate and cadmium(II) nitrate tetrahydrate, respectively, with tr_2ad in the 1:1 molar ratio (methanol, 100 °C, 24 h). All the compounds were isolated, in reasonable yields (55–70%), in the form of air- and moisture-stable microcrystalline powders, insoluble in water and in most common organic solvents (see Section 3.2).

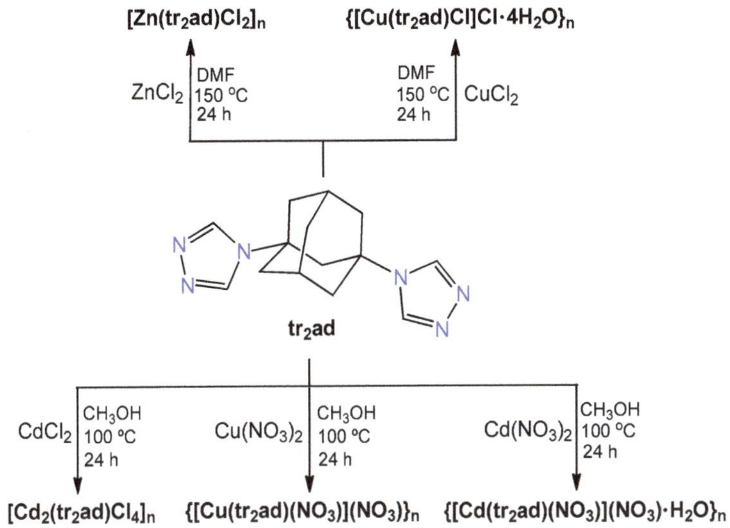

Scheme 2. Synthetic paths for the formation of the tr_2ad-based coordination polymers (CPs) described in this work.

The IR spectrum of the tr_2ad ligand (Figure S1, Supplementary Materials) shows a strong absorption band at 1517 cm^{-1}, which is assigned to the stretching vibration of the triazolyl ring [36]. In the case of the title CPs, this absorption is shifted towards higher wavenumbers (1551–1539 cm^{-1}), as a consequence of the ligand coordination to the metal ions (Figure 1). The medium-intensity broad bands centered around 3400 cm^{-1} in the IR spectra of compounds $\{[Cu(tr_2ad)Cl]Cl{\cdot}4H_2O\}_n$ and $\{[Cd(tr_2ad)(NO_3)](NO_3){\cdot}H_2O\}_n$ (Figure 1), characteristic of the O–H stretching vibration, witness the presence of water molecules.

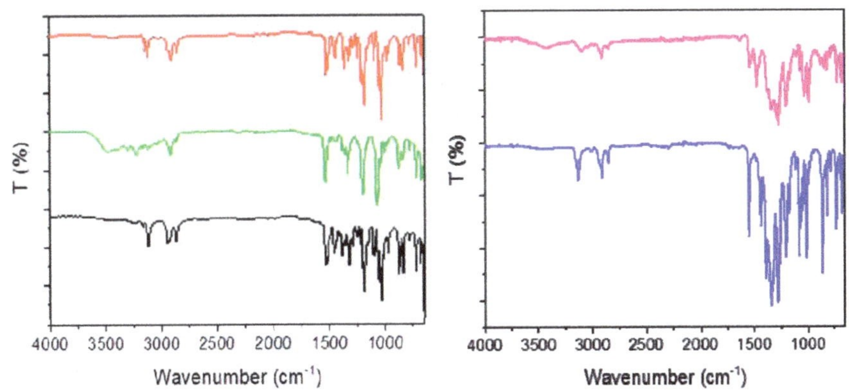

Figure 1. IR spectra of $[Zn(tr_2ad)Cl_2]_n$ (black), $\{[Cu(tr_2ad)Cl]Cl{\cdot}4H_2O\}_n$ (green), $[Cd_2(tr_2ad)Cl_4]_n$ (red), $\{[Cu(tr_2ad)(NO_3)](NO_3)\}_n$ (blue) and $\{[Cd(tr_2ad)(NO_3)](NO_3){\cdot}H_2O\}_n$ (fuchsia).

A deeper analysis of the IR spectra of $\{[Cu(tr_2ad)(NO_3)](NO_3)\}_n$ and $\{[Cd(tr_2ad)(NO_3)](NO_3){\cdot}H_2O\}_n$ (Figure 1) allows to differentiate among the uncoordinated and coordinated forms of the nitrate anion. Indeed, the strong bands located at 1439 and 1282 cm^{-1} for $\{[Cu(tr_2ad)(NO_3)](NO_3)\}_n$, and at 1478 and 1275 cm^{-1} for $\{[Cd(tr_2ad)(NO_3)](NO_3){\cdot}H_2O\}_n$, assigned to the asymmetric and symmetric stretching vibrations of the nitrate group, together with the presence of two very weak bands, at 1755 and

1733 cm^{-1} for {[Cu(tr$_2$ad)(NO$_3$)](NO$_3$)}$_n$ and at 1748 and 1717 cm^{-1} for {[Cd(tr$_2$ad)(NO$_3$)](NO$_3$)·H$_2$O}$_n$, suggest the presence of μ_2:η^2 nitrate anions [37,38]. At variance, the bands observed at 1394 and 1346 cm^{-1} for {[Cu(tr$_2$ad)(NO$_3$)](NO$_3$)}$_n$ and at 1374 and 1339 cm^{-1} for {[Cd(tr$_2$ad)(NO$_3$)](NO$_3$)·H$_2$O}$_n$, together with the band centered at 1073 cm^{-1} in {[Cu(tr$_2$ad)(NO$_3$)](NO$_3$)}$_n$ and at 1076 cm^{-1} in {[Cd(tr$_2$ad)(NO$_3$)](NO$_3$)·H$_2$O}$_n$, can be ascribed to the asymmetric and symmetric stretching modes of uncoordinated nitrate anions [39].

2.2. Thermal Behaviour

Thermogravimetric analyses (TGAs) were performed on the five compounds from 30 °C to 700 °C under a flow of nitrogen. The resulting TGA curves are gathered in Figure 2. Compound [Zn(tr$_2$ad)Cl$_2$]$_n$ is stable up to 350 °C, temperature at which a slow decomposition begins. In the temperature range 30–120 °C, {[Cu(tr$_2$ad)Cl]Cl·4H$_2$O}$_n$ undergoes a weight loss of ca. 15%, which reasonably corresponds to the evolution of four water molecules per formula unit (calculated weight loss 15.1%).

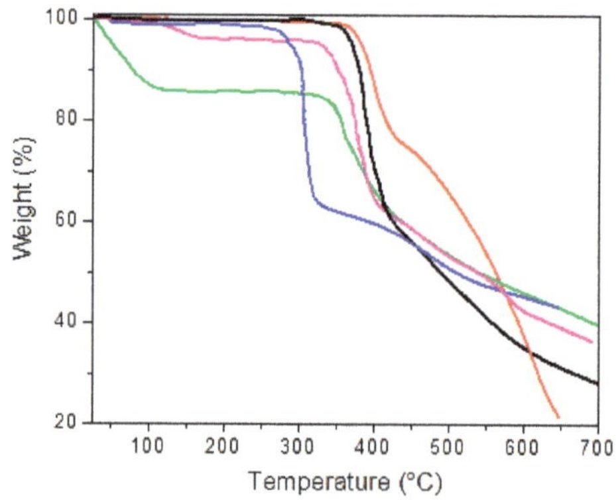

Figure 2. Thermogravimetric analysis (TGA) traces of [Zn(tr$_2$ad)Cl$_2$]$_n$ (black), {[Cu(tr$_2$ad)Cl] Cl·4H$_2$O}$_n$ (green), [Cd$_2$(tr$_2$ad)Cl$_4$]$_n$ (red), {[Cu(tr$_2$ad)(NO$_3$)](NO$_3$)}$_n$ (blue) and {[Cd(tr$_2$ad)(NO$_3$)] (NO$_3$)·H$_2$O}$_n$ (fuchsia).

After solvent loss, no further weight loss is observed up to the decomposition onset at 325 °C. Upon heating, compound {[Cu(tr$_2$ad)(NO$_3$)](NO$_3$)}$_n$ does not undergo any weight loss up to 250 °C, the temperature at which decomposition starts. To the best of our knowledge, the only known 2-D coordination polymers containing the tr$_2$ad ligand of which the thermal behavior have been studied are [Cu$^{II}_2$(tr$_2$ad)$_4$](Mo$_8$O$_{26}$), [Cu$^{II}_4$(μ_4-O)(tr$_2$ad)$_2$(MoO$_4$)$_3$]·7.5H$_2$O [40] and [Cu$_3$(tr$_2$ad)$_4$(H$_2$O)$_4$](SiF$_6$)$_3$·16H$_2$O [41], which decompose at 310 °C, 240 °C and 190 °C, respectively. Compound [Cd$_2$(tr$_2$ad)Cl$_4$]$_n$ displays the highest thermal robustness, peaking up to 365 °C. Until this temperature, no weight loss is observed. Finally, compound {[Cd(tr$_2$ad)(NO$_3$)](NO$_3$)·H$_2$O}$_n$ undergoes a weight loss of ca. 3.5% in the range 30–150 °C, which reasonably corresponds to the release of one water molecule per formula unit (calculated weight loss 3.4%). After this event, no further weight losses are observed up to the decomposition onset at 330 °C. To the best of our knowledge, in no case the thermal behavior of the known Cd^{II} 2-D coordination polymers containing the tr$_2$ad ligand has been investigated, so that a comparison cannot be carried out. For the title compounds, at the end of the heating process, black residues, possibly containing carbonaceous species, have been recovered.

2.3. Crystal and Molecular Structures

Tr$_2$ad crystallizes in the monoclinic space group $P2_1/n$. The asymmetric unit contains one tr$_2$ad molecule in general position. Figure S2a shows the Ortep drawing at 40% probability level. Due to the lack of conventional hydrogen-bond donors, the crystal structure of tr_2ad only features a network of weak CH···N interactions, with shortest C···N distances of 3.350(2) Å. Both triazole and adamantane CH groups act as unconventional hydrogen bond donors, and most of these non-bonding interactions are directional. Two pairs of such CH···N interactions, together with a slipped π/π interaction among adjacent triazole rings (centroid-centroid distance 3.82 Å, slippage angle 6.6°), concur to the formation of tr_2ad centrosymmetric dimers (Figure 3a). Such self-association is reminiscent of the pairing of 1,3,5-triphenyladamantane molecules prompted by weak CH···π interactions [42].

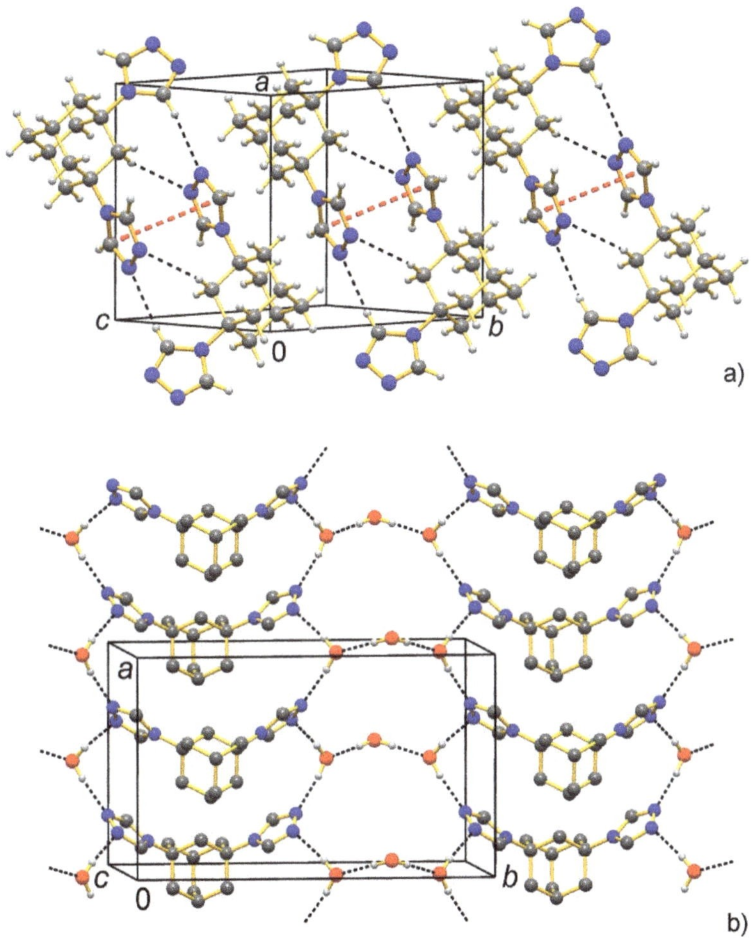

Figure 3. Representation of portion of the crystal structure of (**a**) tr_2ad and (**b**) $tr_2ad{\cdot}3H_2O$, showing the principal supramolecular motifs created by non-bonding interactions involving the triazole nitrogen atoms as acceptors: (**a**) multiple CH···N interactions (black dashed lines) concur with π/π stacking interactions (red dashed lines) to form tr_2ad dimers; (**b**) OH···N and OH···O hydrogen bonds (black dashed lines) support the formation of 2-D supramolecular layers. Atoms colour code: C, grey; H, light grey; N, blue; O, red.

The hydrate ligand $tr_2ad{\cdot}3H_2O$ crystallizes in the orthorhombic space group *Pnma*. The asymmetric unit contains half of a tr_2ad molecule and half of a H_2O molecule, both situated across a mirror plane (Wyckoff letter *c*), and one water molecule in general position. Figure S2b shows the Ortep drawing at 30% probability level. The primary intermolecular interactions in the crystal structure are conventional

OH···N hydrogen bonds (O···N = 2.896(3), 2.932(3) Å) involving all the triazole nitrogen atoms as acceptors. These interactions assemble the tr_2ad and water molecules (in a 1:2 ratio) into 1-D strips along the crystallographic *a*-axis (Figure 3b). Additional water molecules establish bridges between the strips through pairs of symmetry-equivalent OH···O bonds (O···O = 2.764(2) Å) (Figure 3b). The 2-D hydrogen-bond connectivity comprises water trimers H_2O···H–O–H···OH_2 linked to four triazole-N sites. Overall, the crystal structures of tr_2ad and $tr_2ad{\cdot}3H_2O$ reveal the potentiality of triazole-N^1,N^2 atoms as efficient hydrogen-bond acceptors.

Compound $[Zn(tr_2ad)Cl_2]_n$ crystallizes in the orthorhombic space group $P2_12_12_1$. The asymmetric unit contains one Zn^{II} ion, one tr_2ad ligand and two chloride anions, all in general positions. The metal centre shows a $ZnCl_2N_2$ tetrahedral stereochemistry (Figure 4a; the Figure caption collects the values of the bond distances and angles at the metal ion), defined by two chloride anions and the nitrogen atoms of the triazole rings of two tr_2ad ligands. The ligands are exo-bidentate (μ_2-κN^1:κ$N^{1'}$) and bridge neighbouring Zn^{II} ions along 1-D polymeric chains (Figure 4b) of pitch 11.120(4) Å parallel to the [001] crystallographic direction (this occurrence rationalizing the preferred orientation pole; see Section 3.3). The chains pack in the *ab* plane defining a rectangular motif (Figure 4c). Non-bonding interactions of the kind C–H···N (C···N 3.2 Å) and C–H···Cl involving both chloride anions (C···Cl 3.5–3.7 Å) are at work within the chains and between nearby chains, respectively. No empty volume is present [43].

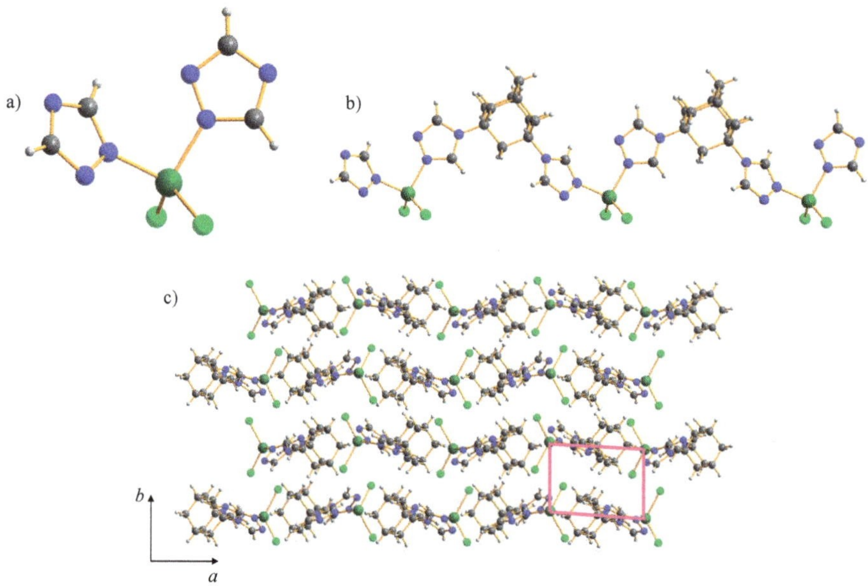

Figure 4. Representation of the crystal structure of $[Zn(tr_2ad)Cl_2]_n$: (**a**) the coordination sphere of the Zn^{II} ions; (**b**) portion of the 1-D polymeric motif running along the [001] crystallographic direction; (**c**) portion of the packing, viewed along the [001] crystallographic direction. Horizontal axis, *a*; Vertical axis, *b*. Highlighted in fuchsia the rectangular packing of the 1-D chains. Atoms colour code: C, grey; H, light grey; Cl, light green; N, blue; Zn, green. Main bond distances (Å) and angles (°) at the metal ions: Zn–Cl 2.153(7), 2.316(7); Zn–N 1.944(9), 2.01(1); shortest intra-chain Zn···Zn 11.120(3); N–Zn–N 99.5(5); Cl–Zn–Cl 117.7(2); Cl–Zn–N 107.0(6), 107.1(5), 110.0(5), 113.7(7).

Compound $\{[Cu(tr_2ad)Cl]Cl{\cdot}4H_2O\}_n$ crystallizes in the monoclinic space group $P2_1/m$. The asymmetric unit is composed by half of a metal centre (on an inversion centre, Wyckoff letter *b*), half of a tr_2ad ligand, two halves of a chloride anion and two halves of a water molecule (all on mirror planes, Wyckoff letter *e*), as well as one water molecule (in general position). The Cu^{II} ions are hexa-coordinated in *trans*-$CuCl_2N_4$ octahedral geometry defined by the nitrogen atoms of four tr_2ad ligands and one of the two independent chloride anions (Figure 5a; the main bond distances and angles at the metal ions are reported in the Figure caption). μ-coordination by triazole rings and coordinated chloride anions brings about the formation of 1-D helices of metal ions (Figure 5b) with pitch 3.5863(2) Å (half of the *b*-axis) running along the crystallographic direction [010]. The tr_2ad ligands, which are overall exo-tetradentate (μ_4-κN^1:κN^2:κ$N^{1'}$:κ$N^{2'}$), connect the helices along the crystallographic direction [001], bringing about the formation of 2-D double-layers parallel to the *bc* crystallographic plane (Figure 5c). The layers pack, staggered, along the *a*-axis. The second independent chloride anion and one of the four independent water molecules occupy the rhombic cavities formed by the ligands within the double-layers (Figure 5c) and are involved in a HO–H···Cl non-bonding interaction (O···Cl 2.58(3) Å; Figure S7a). The other three water molecules are located in the inter-layer space and, by means of hydrogen bonds (O···O 2.44(2), 2.91(2), 3.07(2) Å; Figure S3a), define a 1-D supramolecular chain running parallel to the *b*-axis. Finally, the double layers are reinforced by C–H···Cl interactions (C···Cl 3.3–3.4 Å; Figure S3b). No empty volume is envisaged [43].

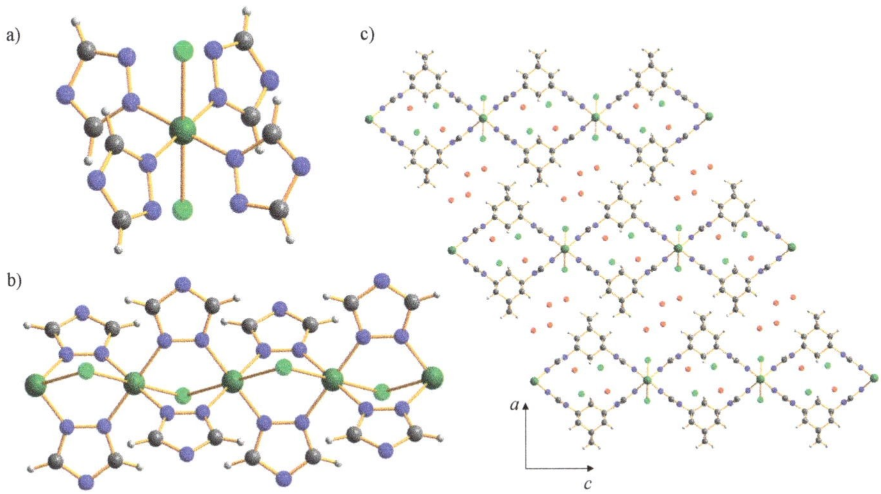

Figure 5. Representation of the crystal structure of $\{[Cu(tr_2ad)Cl]Cl{\cdot}4H_2O\}_n$: (**a**) the coordination sphere of the metal ions; (**b**) portion of the 1-D helix; (**c**) portion of the packing, viewed along the [010] crystallographic direction. Horizontal axis, *c*; vertical axis, *a*. For the non-bonding interactions quoted in the text the reader is addressed to Figure S3. Atoms colour code: C, grey; H, light grey; Cl, light green; Cu, green; N, blue; O, red. Main bond distances (Å) and angles (°) at the metal ions: Cu–Cl 2.632(7); Cu–N 1.969(4), 2.009(6); shortest Cu···Cu intra-chain 3.5863(2); shortest Cu···Cu inter-chain 10.9798(7); N–Cu–N 80.4(3), 99.6(3), 180; Cl–Cu–Cl 180; Cl–Cu–N 88.1(3), 89.7(3), 90.3(3), 91.9(3).

Compound $[Cd_2(tr_2ad)Cl_4]_n$ crystallizes in the triclinic space group *P*-1. The asymmetric unit equals the formula unit, i.e., it contains two cadmium(II) ions, four chloride anions and one tr_2ad ligand, all in general positions. Both independent metal centres are hexa-coordinated and show an octahedral stereochemistry, though of different kind, namely: *cis*-CdN_2Cl_4 and $CdNCl_5$ (Figure 6a; the main bond distances and angles at the metal ions are reported in the Figure caption). Three of the four chloride anions bridge adjacent metal centres, while the fourth one behaves as a terminal ligand. The tr_2ad spacer is exo-tridentate (μ_3-κN^1:κN^2:κ$N^{1'}$). The reciprocal disposition of cations and anions brings about the formation of 1-D polymeric strands (Figure 6b) running along the [100] crystallographic direction. The tr_2ad linkers bridge nearby strands leading to the formation of 2-D double-layers parallel to the (01-1) plane and packing, staggered, along the [011] direction (Figure 6c). The reciprocal disposition of the spacers within a layer brings about the formation of intra-layer rhombic cavities, in which the terminal chloride anions are directed (Figure 6c). This structural motif is analogous to that found in {[Cu(tr$_2$ad)Cl]Cl·4H$_2$O}$_n$ (see above). Weak intra- and inter-layer C–H···Cl interactions (C···Cl 3.3–3.7 Å) are present. No empty volume is observed [43].

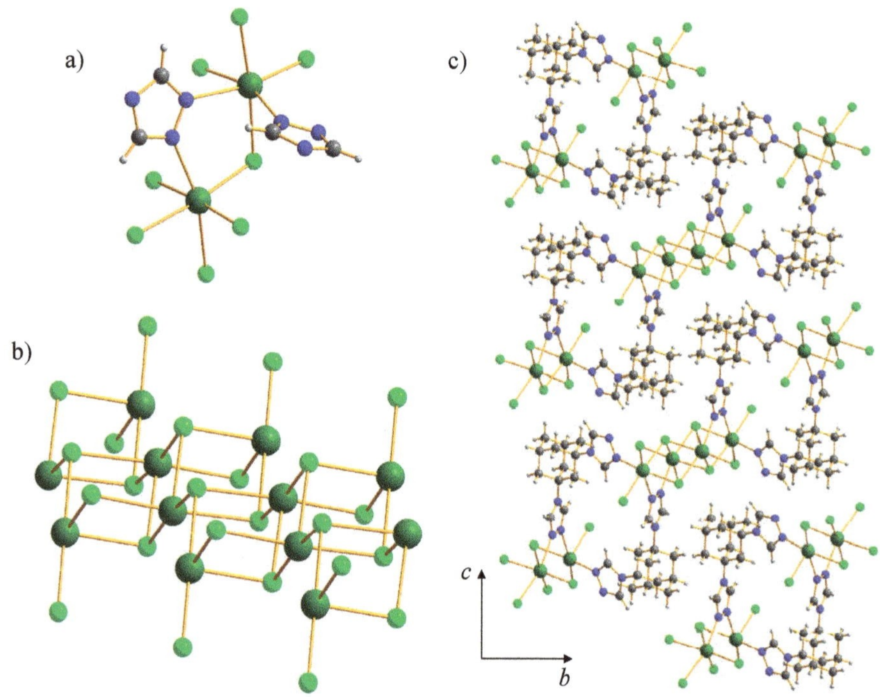

Figure 6. Representation of the crystal structure of [Cd$_2$(tr$_2$ad)Cl$_4$]$_n$: (**a**) the coordination sphere of the metal ions; (**b**) portion of the 1-D polymeric strands; (**c**) portion of the packing, viewed along the [100] crystallographic direction. Horizontal axis, *b*; vertical axis, *c*. Atoms colour code: C, grey; H, light grey; Cl, light green; Cd, green; N, blue. Main bond distances (Å) and angles (°) at the metal ions: Cd1–Cl 2.59(2), 2.62(2), 2.75(2), 2.82(2), 2.87(2); Cd1–N 2.42(3); Cd2–Cl 2.58(2), 2.59(1), 2.62(2), 2.78(2); Cd2–N 2.32(2), 2.32(3); shortest intra-strand Cd···Cd 3.842(8)–4.241(8); shortest inter-strand Cd···Cd 7.526(7) Cl–Cd1–N 84(1)–165.0(1); Cl–Cd1–Cl 82.5(5)–174.2(5); Cl–Cd2–N 77(1)–165(1); Cl–Cd2–Cl 85.6(5)–173.5(6); N–Cd2–N 92.2(8).

Compound {[Cu(tr$_2$ad)(NO$_3$)](NO$_3$)}$_n$ crystallizes in the orthorhombic space group *Pnma*. The asymmetric unit is composed by one Cu^{II} ion, two halves of nitrate anions and half of a tr$_2$ad spacer, all lying on mirror planes (Wyckoff letter *h*). The metal centre is hexa-coordinated in *trans*-CuN$_4$O$_2$ stereochemistry (Figure 7a; the main bond distances and angles at the metal ions are reported in the Figure caption), defined by four tr$_2$ad linkers and one of the two independent nitrate anions. The latter bridges (μ_2-κO^1:κO^2) nearby metal centres 3.54(2) Å apart, while the other nitrate anion is not coordinated. The tr$_2$ad ligand is exo-tetradentate (μ_4-κN^1:κN^2:$\kappa N^{1'}$:$\kappa N^{2'}$). μ_2-bridging of the nitrate anions and triazole rings brings about the formation of 1-D chains running along the [100] direction (Figure 7b).

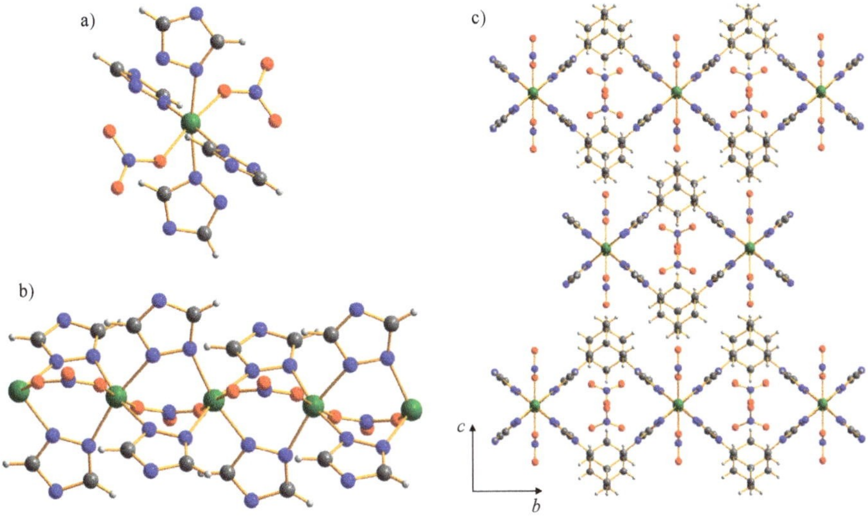

Figure 7. Representation of the crystal structure of {[Cu(tr$_2$ad)(NO$_3$)](NO$_3$)}$_n$: (**a**) the coordination sphere of the metal ions; (**b**) portion of the 1-D polymeric chain; (**c**) portion of the packing, viewed along the [100] crystallographic direction. Horizontal axis, *b*; vertical axis, *c*. Atoms colour code: C, grey; H, light grey; Cu, green; N, blue; O, red. Main bond distances (Å) and angles (°) at the metal ions: Cu–N 2.03(1), 2.148(9); Cu–O 2.09(2), 2.26(2); intra-chain shortest Cu···Cu 3.54(2); inter-chain shortest Cu···Cu 10.7226(7); N–Cu–N 76.4(6), 97.01(3), 173.5(5); O–Cu–O 180(1); O–Cu–N 68.9(5), 70.3(6), 109.8(6), 111.0(6).

Nearby chains are connected along [010] by the tr$_2$ad spacers within 2-D polymeric double-layers parallel to the (001) plane and packing, staggered, along the [001] direction (Figure 7c; this occurrence explains the preferred orientation pole—see Section 3.3). The reciprocal disposition of the tr$_2$ad linkers within a layer brings about the formation of intra-layer rhombic cavities, in which the not coordinated nitrate anions are located (Figure 7c) and involved in C–H···O non-bonding interactions (C···O 2.6–3.2 Å). The structural motif is analogous to that found in the Cu^{II} and Cd^{II} compounds described above. No empty volume is observed [43].

Compound {[Cd(tr$_2$ad)(NO$_3$)](NO$_3$)·H$_2$O}$_n$ crystallizes in the monoclinic space group *C*2/*c*. The asymmetric unit contains two halves of Cd^{II} ions (one on an inversion centre, Wyckoff position *b*, the other one on a two-fold axis, Wyckoff position *e*), two nitrate anions, one tr$_2$ad ligand and one water molecule—all in general positions. The metal centres are hexa-coordinated in *trans*-CdN$_4$O$_2$ octahedral stereochemistry defined by four ligands and two nitrate anions (Figure 8a; the main bond distances and angles at the metal ions are reported in the Figure caption). One of the two independent nitrate anions bridge (μ_2-κO^1:κO^2) neighbouring metal centres, while the other one is not coordinated. The tr$_2$ad spacer is exo-tetradentate (μ_4-κN^1:κN^2:$\kappa N^{1'}$:$\kappa N^{2'}$). μ_2-Coordination of the nitrate anions and triazole rings is responsible for the formation of 1-D chains parallel to the [001] crystallographic direction (Figure 8b). Adjacent chains are bridged along the [010] direction to yield 2-D double-layers parallel to the *bc* plane and packing, staggered, along the *a*-axis (Figure 8c; this occurrence explains the preferred orientation pole—see Section 3.3). The reciprocal disposition of the tr$_2$ad linkers within a layer brings about the formation of intra-layer rhombic cavities, in which the not coordinated nitrate anions are located (Figure 8c). The structural motif is analogous to that found in the Cu^{II} and Cd^{II} compounds described above. The water molecules are located in the inter-layer space (Figure 8c) and are involved in C–H···O non-bonding interactions (C···O 2.9–3.2 Å) with adjacent tr$_2$ad ligands. No empty volume is observed [43].

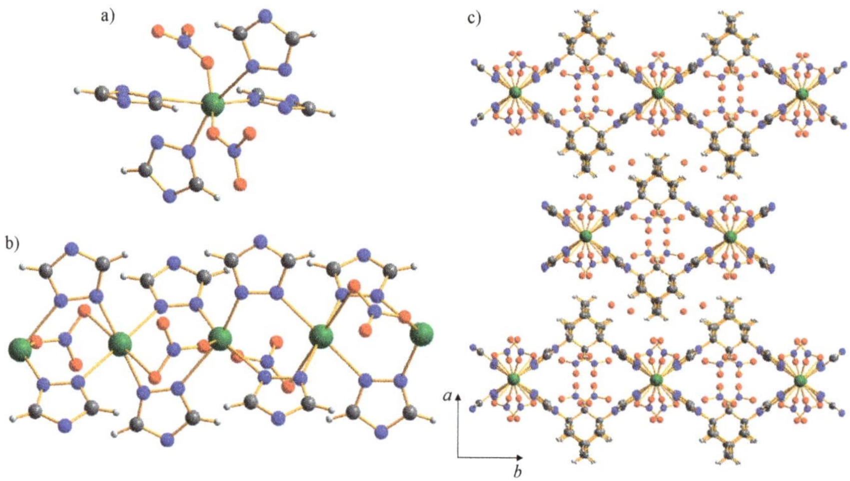

Figure 8. Representation of the crystal structure of {[Cd(tr$_2$ad)(NO$_3$)](NO$_3$)·H$_2$O}$_n$: (**a**) the coordination sphere of the metal ions; (**b**) portion of the 1-D polymeric chains; (**c**) portion of the packing, viewed in perspective along the [001] crystallographic direction. Horizontal axis, *b*; vertical axis, *a*. Atoms colour code: C, grey; H, light grey; Cu, green; N, blue; O, red. Main bond distances (Å) and angles (°) at the metal ions: Cd1–N 2.24(1), 2.30(1); Cd1–O 1.54(3); Cd2–N 2.17(1), 2.23(1); Cd2–O 2.79(4); intra-chain Cd···Cd 3.8724(3); inter-chain Cd···Cd 11.39(2); N–Cd1–N 80.2(5)–176.3(5); O–Cd1–O 168(3); N–Cd1–O 72(1)–105(1); N–Cd2–N 88.7(6), 91.3(6), 180; O–Cd2–O 180; N–Cd2–O 68(1)–110.1(8).

2.4. Comparative Structure Analysis

A search in the Cambridge Structural Database (v 2020.1) for coordination compounds containing the tr$_2$ad ligand has revealed the existence of 22 coordination polymers. Table 1 collects key structural aspects (coordination sphere and geometry at the metal ion, tr$_2$ad ligand hapticity, polymer dimensionality) of these compounds. The following observations can be carried out:

1. Among the CPs retrieved in the literature (Table 1), 14 contain Cu^{II} or Cd^{II}, while the others feature Cu^{I}, Ag^{I}, Mo^{II} or Fe^{II}. Hence, to the best of our knowledge, [Zn(tr$_2$ad)Cl$_2$]$_n$ is the first example of Zn^{II}-based coordination compound containing the tr$_2$ad ligand.
2. As regards the stereochemistry at the metal ion, apart from [Cd$_3$(tr$_2$ad)$_2$I$_6$], in which one of the two independent Cd^{II} ions shows a tetrahedral geometry, in all the known Cd^{II} CPs the metal centre adopts an octahedral geometry, as in [Cd(tr$_2$ad)Cl$_4$]$_n$ and {[Cd(tr$_2$ad)(NO$_3$)](NO$_3$)·H$_2$O}$_n$. On the other hand, while in {[Cu(tr$_2$ad)Cl]Cl·4H$_2$O}$_n$ and {[Cu(tr$_2$ad)(NO$_3$)](NO$_3$)}$_n$ the metal ion is in octahedral stereochemistry, in the known CPs the Cu^{II} coordination number varies in the range 4–6, with different coordination geometries associated (Table 1).
3. Upon comparing the values of the M–N distances (M = Cu^{II} or Cd^{II}, N = tr$_2$ad nitrogen atom), it appears (Figure S4) that the novel materials share distances comparable with those of the literature CPs.
4. Apart from [Zn(tr$_2$ad)Cl$_2$]$_n$, all the other compounds studied in this work are 2-D coordination polymers characterized by the same structural motif (see Figures 4–8). At variance, Table 1 shows that in the known compounds the dimensionality ranges from 1-D to 3-D. Interestingly, the structural motif observed in the title Cu^{II} and Cd^{II} derivatives, with rhombic cavities within 2-D strands, is shown also by COVFIE, COVFOK, KEHDEI, KEMLEV and TUGSIY, containing Cu^{II} or Cd^{II} ions, and UZAKIQ, containing the Mo^{II} ion.

Table 1. Main structural properties of the known coordination polymers containing the tr_2ad ligand. Abbreviations: M = metal ion; OC = octahedral; SP = square planar; SQP = square pyramidal; TB = trigonal bipyramidal; TD = tetrahedral; TP = trigonal planar; H_3btc = 1,3,5-benzenetricarboxylic acid; H_4adtc = 1,3,5,7-adamantane-tetracarboxylic acid; [a] M–F.

CSD Code	Molecular Formula	M Stereochemistry	M Geometry	M–N (Å)	M–X (Å)	Tr_2ad Apticity	Dim.	Ref.
COVFIE	$[Cu_2(tr_2ad)_4](Mo_8O_{26})$	CuN_5	SP	1.968–2.247	-	Exo-bidentate Exo-tridentate	2-D	[40]
COVFOK	$[Cu_4(tr_2ad)_2(\mu_4\text{-}O)(MoO_4)_3]\cdot 7.5H_2O$	CuN_2O_3 CuN_2O_3 CuN_2O_3	TB TB TB	2.078, 2.093 2.005 2.009	1.923–2.028 1.937–2.090 1.937–2.067	Exo-tetradentate (2×)	2-D	[40]
COVFUQ	$[Cu_2(tr_2ad)_2](Mo_2O_7)\cdot H_2O$	CuN_3 CuN_3	TP TP	1.928–2.002 1.934–2.000	- -	Exo-tridentate (2×)	1-D	[40]
KEMLEV	$[Cu(tr_2ad)(SO_4)]\cdot 3H_2O$	*trans*-CuN_4O_2	OC	2.009–2.012	2.406	Exo-tetradentate	2-D	[41]
KEMLIZ	$[Cu_3(tr_2ad)_4(H_2O)_2(SO_4)_2](SO_4)\cdot 28H_2O$	CuN_6 CuN_4O	OC SQP	2.024–2.365 1.995–2.223	- 1.958	Exo-tetradentate Exo-tridentate	2-D	[41]
KEMLOF	$[Cu_3(tr_2ad)_4(H_2O)_2](SiF_6)_3\cdot 16H_2O$	CuN_6 *cis*-CuN_4O_2	OC OC	2.001-2.413 1.988–2.269	- 1.987, 2.554	Exo-tetradentate Exo-tridentate	2-D	[41]
PODMAX	$[Cu_4(tr_2ad)_2(btc)_2(\mu_3\text{-}OH)_2]$	$CuNO_4$ CuN_2O_3	SQP TB	2.411 1.989–2.010	1.915–1.999 1.953–2.165	Exo-tridentate	3-D	[44]
SERCUP	$[Cu_4(tr_2ad)_2(H\text{-}adtc)_2(OH)_2(H_2O)_2]\cdot 4H_2O$	$CuNO_5$ *cis*-CuN_2O_4 CuN_2O_3 $CuNO_4$	OC OC SQP SQP	1.981 1.989–2.334 2.005–2.271 1.987	1.905–2.685 1.957–2.691 1.950–1.996 1.923–2.279	Exo-tridentate	2-D	[45]
TUGSUK	$[Cu_3(tr_2ad)_4(SO_4)(H_2O)_3](SO_4)_2\cdot 34H_2O$	CuN_6 *cis*-CuN_4O_2 CuN_4O	OC OC SQP	1.989–2.295 1.990–2.250 1.957–2.291	- 1.989, 2.408 1.989	Exo-tetradentate (2×) Exo-tridentate (2×)	3-D	[46]
TUGTAR	$[Cu_2(\mu\text{-}OH)(tr_2ad)_2](NO_3)_3\cdot 4H_2O$	CuN_6 *cis*-CuN_4O_2 *trans*-CuN_2O_4	OC OC OC	2.026–2.527 1.982–2.212 2.000	- 1.921–2.788 1.909–2.686	Exo-tetradentate (2×)	3-D	[46]
ILUFEB	$[Cd_3(tr_2ad)_3](SeCN)_6$	CdN_6 CdN_6	OC OC	2.289–2.370 2.332–2.353	- -	Exo-tridentate (4×)	3-D	[47]
KEHDEI	$[Cd_2(tr_2ad)_2(H_2O)_4](CdBr_4)_2$	*cis*-CdN_4O_2 *cis*-CdN_4O_2 $CdBr_4$	OC OC TD	2.266–2.343 2.273–2.323 -	2.377–2.382 2.370–2.406 2.567–2.629	Exo-bidentate (4×)	3-D	[48]

Table 1. *Cont.*

CSD Code	Molecular Formula	M Stereochemistry	M Geometry	M–N (Å)	M–X (Å)	Tr_2ad Apticity	Dim.	Ref.
KEHDIM	$[Cd_3(tr_2ad)_2I_6]$	CdN_6 $CdNI_3$	OC TD	2.305–2.403 2.283	- 2.744–2.756	Exo-bidentate (2×)	2-D	[48]
TUGSIY	$[Cd_3(tr_2ad)_3(\mu\text{-}NCS)_3](NCS)_3$	$CdN_{1-x}S_x$ $(0 < x < 1)$ CdN_6	OC OC	2.296–2.473 2.312–2.405	2.662	Exo-tetradentate (3×)	2-D	[46]
TUGSOE	$[Cd_3(tr_2ad)_6](NO_3)_6{\cdot}22H_2O$	CdN_6 CdN_6	OC OC	2.324–2.356 2.297–2.381	-	Exo-tridentate (3×)	3-D	[46]
HUSQES	$[Mo_2(tr_2ad)O_6)]{\cdot}6H_2O$	*cis*-MoN_2O_4	OC	2.366, 2.375	1.702–1.932	Exo-tetradentate	3-D	[49]
LUYRII	$[Mo_2(tr_2ad)O_6)]{\cdot}H_2O$	*cis*-MoN_2O_4	OC	2.405	1.714–1.910	Exo-tetradentate	2-D	[50]
UZAKIQ	$[Mo_2(tr_2ad)F_2O_5]$	MoN_2O_3F MoN_2O_3F	OC OC	2.399 2.380	1.710–1.918 1.917 [a] 1.706–1.914 1.934 [a]	Exo-tetradentate	1-D/2-D	[51]
VEPDEB	$\{Fe_3(tr_2ad)_4[Au(CN)_2]_2\}[Au(CN)_2]_4{\cdot}8H_2O$	FeN_6 FeN_6	OC OC	1.910–1.970 1.975–1.987	- -	Exo-tetradentate (2×) Exo-tridentate	3-D	[52]
KEHDOS	$[Ag(tr_2ad)](NO_3){\cdot}H_2O$	AgN_4 AgN_4	TD TD	2.185–2.502 2.185–2.528	- -	Exo-tetradentate (2×)	3-D	[48]
WEJWAL	$[Ag(tr_2ad)](ClO_4)$	AgN_3	TP	2.191–2.347	-	Exo-tridentate	2-D	[53]
WEJWEP	$[Ag_2(tr_2ad)_2(VO_2F_2)_2]{\cdot}H_2O$	AgN_3O	TD	2.209–2.464	2.558	Exo-tetradentate	1-D	[53]

3. Materials and Methods

3.1. General

All reagents and solvents were purchased from Sigma-Aldrich (Darmstadt, Germany) and used as received, without further purification. The ligand 1,3-bis(1,2,4-triazol-4-yl)adamantane (tr_2ad) was synthesized by the acid-catalyzed condensation reaction of 1,3-diaminoadamantane and *N*,*N*-dimethylformamide azine, according to an already reported method [46]. A detailed description regarding the preparation of the intermediates (Scheme S1), on which no details have ever been reported before, is provided in the Supplementary Materials. NMR spectra (DMSO-d_6, δ, ppm) were recorded on a Bruker 400 MHz spectrometer. The IR spectra were recorded from 4000 to 650 cm^{-1} with a PerkinElmer Spectrum 100 instrument (Perkin-Elmer, Shelton, CT, USA) by attenuated total reflectance on a CdSe crystal. Elemental analyses (carbon, hydrogen, and nitrogen %) were performed with a Fisons Instruments 1108 CHNS-O elemental analyzer (Thermo Scientific, Waltham, MA, USA). Before the analytical characterization was carried out, all the samples were dried under vacuum (50 °C, ~0.1 Torr) until a constant weight was reached. Thermogravimetric analyses (TGAs) were carried out under a N_2 flow (25 mL/min), in the temperature range 30–700 °C and with a heating rate of 5 °C/min, using a PerkinElmer STA 6000 simultaneous thermal analyzer (Perkin-Elmer, Shelton, CT, USA).

3.2. Synthesis of the Tr_2ad-Based CPs

3.2.1. Synthesis of $[Zn(tr_2ad)Cl_2]_n$

Tr_2ad (0.054 g, 0.2 mmol) was dissolved in *N*,*N*-dimethylformamide (DMF) (5 mL) and the obtained solution was left under stirring at room temperature for 5 min. Then, $ZnCl_2{\cdot}2H_2O$ (0.017 g, 0.1 mmol) was added, and the resulting solution was introduced into a high-pressure glass tube and heated at 150 °C for 24 h. Slow cooling of the solution to room temperature, followed by partial slow evaporation of the solvent, afforded a white solid which was filtered off, washed twice with DMF, dried under vacuum and identified as $[Zn(tr_2ad)Cl_2]$. Yield: 55%. $[Zn(tr_2ad)Cl_2]$ is insoluble in alcohols, acetone, acetonitrile, chlorinated solvents, DMF, dimethylsulfoxide (DMSO) and water. Elem. anal. calc. for $C_{14}H_{18}Cl_2N_6Zn$ (FW = 406.65 g/mol): C, 41.35; H, 4.46; N, 20.67%. Found: C, 40.95; H, 4.23; N, 20.35%. IR (cm^{-1}): 3160(w), 3110(m) ν(C–$H_{aromatic}$), 2930(m), 2870(m) ν(C–$H_{aliphatic}$), 1539(s) ν(C=N), 1387(m), 1329(m), 1192(vs), 1110(m), 1029(vs), 973(w), 883(m), 934(m), 786(w), 728(m), 680(m), 656(vs).

3.2.2. Synthesis of $\{[Cu(tr_2ad)Cl]Cl{\cdot}4H_2O\}_n$

Tr_2ad (0.054 g, 0.2 mmol) was dissolved in DMF (5 mL) and the obtained solution was left under stirring at room temperature for 5 min. Then, $CuCl_2$ (0.013 g, 0.1 mmol) was added, and the resulting solution was introduced into a high-pressure glass tube and heated at 150 °C for 24 h. Slow cooling of the solution to room temperature, followed by slow partial evaporation of the solvent, afforded a light blue solid which was filtered off, washed twice with DMF, dried under vacuum and identified as $[Cu(tr_2ad)Cl]Cl{\cdot}4H_2O$. Yield: 65%. $[Cu(tr_2ad)Cl]Cl{\cdot}4H_2O$ is insoluble in alcohols, acetone, acetonitrile, chlorinated solvents, DMF, DMSO and water. Elem. anal. calc. for $C_{14}H_{26}Cl_2CuN_6O_4$ (FW = 476.91 g/mol): C, 35.26; H, 5.49; N, 17.62%. Found: C, 34.85; H, 5.33; N, 17.25%. IR (cm^{-1}): 3500–3200(br) ν(H–O), 3200–3000(w) ν(C–$H_{aromatic}$), 2921(m), 2864(w) ν(C–$H_{aliphatic}$), 1543(s) ν(C=N), 1346(s), 1207(vs), 1086(vs), 1045(w), 1021(w), 883(m), 842(w), 786(w), 728(m), 680(m).

3.2.3. Synthesis of $[Cd_2(tr_2ad)Cl_4]_n$

Tr_2ad (0.054 g, 0.2 mmol) was dissolved in methanol (5 mL) and the obtained solution was left under stirring at room temperature for 5 min. Then, $CdCl_2$ (0.036 g, 0.2 mmol) was added, and the resulting solution was introduced into a high-pressure glass tube and heated at 100 °C for 24 h. The white precipitate which was formed was filtered off, washed three times with hot methanol, dried under vacuum and identified as $[Cd_2(tr_2ad)Cl_4]$. Yield: 65%. $[Cd_2(tr_2ad)Cl_4]$ is insoluble in alcohols, acetone, acetonitrile, chlorinated solvents, DMF, DMSO and water. Elem. anal. calc. for $C_{14}H_{18}Cd_2Cl_4N_6$ (FW = 637.00 g/mol): C, 26.40; H, 2.85; N, 13.19%. Found: C, 26.33; H, 2.92; N, 12.85%. IR (cm^{-1}): 3143(w), 3121(m) ν(C–$H_{aromatic}$), 2915(m), 2858(w) ν(C–$H_{aliphatic}$), 1737(m), 1534(m) ν(C=N), 1371(m), 1253(m), 1191(vs), 1105(m), 1069(s), 1046(vs), 990(m), 883(m), 866(m), 850(m), 791(w), 730(m), 681(m), 658(m).

3.2.4. Synthesis of {[Cu(tr_2ad)(NO_3)](NO_3)}$_n$

Tr_2ad (0.054 g, 0.2 mmol) was dissolved in methanol (5 mL) and the obtained solution was left under stirring at room temperature for 5 min. Then, $Cu(NO_3)_2\cdot 2.5H_2O$ (0.037 g, 0.2 mmol) was added, and the resulting solution was introduced into a high-pressure glass tube and heated at 100 °C for 24 h. A green precipitate was formed, which was filtered off, washed three times with hot methanol, dried under vacuum and identified as [Cu(tr_2ad)(NO_3)](NO_3). Yield: 70%. [Cu(tr_2ad)(NO_3)](NO_3) is insoluble in alcohols, acetone, acetonitrile, chlorinated solvents, DMF, DMSO and water. Elem. Anal. calc. for $C_{14}H_{18}CuN_8O_6$ (FW = 457.95 g/mol): C, 36.72; H, 3.96; N, 24.47%. Found: C, 36.35; H, 3.73; N, 24.19%. IR (cm^{-1}): 3136(m), 3016(vw) ν(C–$H_{aromatic}$), 2916(m), 2856(w) ν(C–$H_{aliphatic}$), 1755(vw), 1733(vw), 1551(s) ν(C=N), 1439(s) ν_{asym}(coordinated NO_3), 1394(vs) ν_{asym}(uncoordinated NO_3), 1346(vs) ν_{sym}(uncoordinated NO_3), 1329(vs), 1282(vs) ν_{sym}(coordinated NO_3), 1210(s), 1180(m), 1119(w), 1089(s), 1073(m), 1054(m), 1019(s), 867(vs), 826(m), 787(w), 738(m), 716(w), 681(m).

3.2.5. Synthesis of {[Cd(tr_2ad)(NO_3)](NO_3)·H_2O}$_n$

Tr_2ad (0.054 g, 0.2 mmol) was dissolved in methanol (5 mL) and the obtained solution was left under stirring at room temperature for 5 min. Then, $Cd(NO_3)_2\cdot 4H_2O$ (0.047 g, 0.2 mmol) was added, and the resulting solution was introduced into a high pressure glass tube and heated at 100 °C for 24 h. The white precipitate which was formed was filtered off, washed three times with hot methanol, dried under vacuum and identified as [Cd(tr_2ad)(NO_3)](NO_3)·H_2O. Yield: 60%. [Cd(tr_2ad)(NO_3)](NO_3)·H_2O is insoluble in alcohols, acetone, acetonitrile, chlorinated solvents, DMF, DMSO and water. Elem. Anal. calc. for $C_{14}H_{20}CdN_8O_7$ (FW = 524.83 g/mol): C, 32.04; H, 3.84; N, 21.35%. Found: C, 31.87; H, 3.55; N, 20.98%. IR (cm^{-1}): 3418(br) ν(H–O), 3098(w), 3043(w) ν(C–$H_{aromatic}$), 2916(m), 2869(w), 2853(w) ν(C–$H_{aliphatic}$), 1748(vw), 1717(vw), 1633(w), 1543(m) ν(C=N), 1478(s) ν_{asym}(coordinated NO_3), 1374(vs) ν_{asym}(uncoordinated NO_3), 1339(vs) ν_{sym}(uncoordinated NO_3), 1301(vs), 1275(vs) ν_{sym}(coordinated NO_3), 1208(vs), 1174(m), 1110(w), 1076(m), 1037(vs), 998(vs), 851(m), 828(m), 789(w), 731(s), 683(s).

3.3. *X-ray Diffraction Structural Analysis*

3.3.1. Structural Analysis of tr_2ad and tr_2ad·3H_2O

The X-ray diffraction data of tr_2ad (colorless prism with dimensions of 0.27 × 0.22 × 0.20 mm) and tr_2ad·3H_2O (colorless prism with dimensions of 0.33 × 0.16 × 0.13 mm) were collected at 173 K on a Bruker APEXII area-detector diffractometer (Bruker, Billerica, MA, USA) equipped with a sealed X-ray tube (Mo-Kα radiation, λ = 0.71073 Å). The data were corrected for Lorentz-polarization effects and for the effects of absorption (multi-scans method). The crystal structures were solved by direct methods and refined against F^2 using the programs SHELXS-97 or SHELXL-2018/1 [54,55]. The non-hydrogen atoms were assigned anisotropic thermal displacement parameters. All the hydrogen atoms were located in difference Fourier maps and then refined freely with isotropic thermal displacement parameters and with soft similarity restraints applied to O–H bond lengths in the structure of tr_2ad·3H_2O.

Crystal data for tr_2ad, FW = 270.34 g mol^{-1}: monoclinic, $P2_1/n$, a = 8.9372(4) Å, b = 8.7877(5) Å, c = 16.8366(8) Å, β = 97.888(2)°, V = 1309.79(11) Å^3, Z = 4, ρ = 1.371 g cm^{-3}, $F(000)$ = 576, $R1$ = 0.039, w$R2$ = 0.095 [$I > 2\sigma(I)$] and $R1$ = 0.053, w$R2$ = 0.1032 (all data) for 2656 data and 253 parameters in the 4.9–52.8° 2θ range. CCDC No. 2034960.

Crystal data for $tr_2ad·3H_2O$, FW = 324.39 g mol^{-1}: orthorhombic, *Pnma*, a = 9.6759(10) Å, b = 16.3052(8) Å, c = 10.0229(11) Å, V = 1581.3(3) $Å^3$, Z = 4, ρ = 1.363 g cm^{-3}, F(000) = 696, $R1$ = 0.049, w$R2$ = 0.087 [$I > 2\sigma(I)$] and $R1$ = 0.110, w$R2$ = 0.104 (all data) for 1656 data and 164 parameters in the 4.8–52.8° 2θ range. CCDC No. 2034961.

3.3.2. Structural Analysis of the Coordination Polymers

Powdered samples (~50 mg) of the five CPs were deposited in the cavity of a silicon free-background sample-holder 0.2 mm deep (Assing S.r.l., Monterotondo, Italy). Powder X-ray diffraction (PXRD) data acquisitions were carried out with a Bruker AXS D8 Advance vertical-scan θ:θ diffractometer (Bruker, Billerica, MA, USA), equipped with a sealed X-ray tube (Cu-Kα, λ = 1.5418 Å), a Bruker Lynxeye linear position-sensitive detector, a filter of nickel in the diffracted beam and the following optical components: primary beam Soller slits (aperture 2.5°), fixed divergence slit (aperture 0.5°), anti-scatter slit (aperture 8 mm). The generator was set at 40 kV and 40 mA. Preliminary PXRD analyses to unveil the purity and crystallinity of the samples were performed in the 2θ range 3.0–35.0°, with steps of 0.02° and time per step of 1 s. PXRD acquisitions for the assessment of the crystal structure were performed in the 2θ range 5.0–105.0°, with steps of 0.02° and time per step of 10 s. After a standard peak search, enabling us to assess the maximum position of the 20–25 lower-angle peaks, indexing was performed applying the Singular Value Decomposition approach [56] implemented in TOPAS-R V3 [57]. The space groups were assigned based on the systematic absences. The crystallographically independent portion of the tr_2ad ligand and nitrate anion were described using rigid bodies built up through the z-matrix formalism, assigning average values to the bond distances and angles (For tr_2ad: $C_{ad/tz}$–C_{ad} = 1.55 Å, C_{tz}–N_{tz}, N_{tz}–N_{tz} = 1.36 Å, C–H = 0.95 Å; Triazole Internal and External Bond Angles = 108 and 126°; Angles at the C_{ad} Atoms = 109.5°. For the Nitrate Anion: N–O = 1.30 Å; O–N–O = 120°.). The structures were solved working in the real space with the Simulated Annealing approach [58], as implemented in TOPAS-R V3. Structures refinement was carried out with the Rietveld method [59], as implemented in TOPAS-R V3. The background was modelled through a polynomial function of the Chebyshev type. An isotropic thermal factor [B_{iso}(M)] was refined for the metal centres; the isotropic thermal factor of lighter atoms was calculated as B_{iso}(L) = B_{iso}(M) + 2.0 ($Å^2$). The peak profile was modelled trough the Fundamental Parameters Approach [60]. The anisotropic shape of the peaks was modelled with the aid of spherical harmonics in all the cases. A correction was applied for preferred orientation adopting the March-Dollase model [61] in the case of $[Zn(tr_2ad)Cl_2]_n$ and $\{[Cu(tr_2ad)(NO_3)](NO_3)\}_n$ (along the [001] pole), as well as of $\{[Cd(tr_2ad)(NO_3)](NO_3)·H_2O\}_n$ (along the [100] pole). The final Rietveld refinement plots are shown in Figures S5–S9.

Crystal data for $[Zn(tr_2ad)Cl_2]_n$, FW = 406.65 g mol^{-1}: orthorhombic, $P2_12_12_1$, a = 14.6240(3) Å, b = 10.1054(2) Å, c = 11.1204(1) Å, V = 1643.40(5) $Å^3$, Z = Z′ = 4, ρ = 1.64 g cm^{-3}, F(000) = 832.0, R_{Bragg} = 0.051, R_p = 0.057 and R_{wp} = 0.078, for 4801 data and 46 parameters in the 9.0–105.0° 2θ range. CCDC No. 2038425.

Crystal data for $\{[Cu(tr_2ad)Cl]Cl·4H_2O\}_n$, FW = 476.91 g mol^{-1}: monoclinic, $P2_1/m$, a = 14.5644(9) Å, b = 7.1726(4) Å, c = 10.9798(6) Å, β = 122.820(3)°, V = 963.9(1) $Å^3$, Z = 4, Z′ = 2, ρ = 1.64 g cm^{-3}, F(000) = 494.0, R_{Bragg} = 0.019, R_p = 0.028 and R_{wp} = 0.040, for 4951 data and 71 parameters in the 6.0–105.0° 2θ range. CCDC No. 2038423.

Crystal data for $[Cd_2(tr_2ad)Cl_4]_n$, FW = 637.00 g mol^{-1}: triclinic, P-1, a = 6.9425(2) Å, b = 12.2352(3) Å, c = 12.6513(2) Å, α = 115.621(1)°, β = 90.837(2)°, γ = 101.165(2)°, V = 944.74(4) $Å^3$, Z = Z′ = 2, ρ = 2.24 g cm^{-3}, F(000) = 616.0, R_{Bragg} = 0.050, R_p = 0.049 and R_{wp} = 0.068, for 4901 data and 68 parameters in the 7.0–105.0° 2θ range. CCDC No. 2038421.

Crystal data for $\{[Cu(tr_2ad)(NO_3)](NO_3)\}_n$, FW = 457.95 g mol^{-1}: orthorhombic, *Pnma*, a = 7.0648(4) Å, b = 10.7226(5) Å, c = 22.495(1) Å, V = 1704.0(2) $Å^3$, Z = 8, Z′ = 4, ρ = 1.79 g cm^{-3}, F(000) = 940.0, R_{Bragg} = 0.025, R_p = 0.037 and R_{wp} = 0.048, for 4951 data and 73 parameters in the 6.0–105.0° 2θ range. CCDC No. 2038424.

Crystal data for $\{[Cd(tr_2ad)(NO_3)](NO_3)·H_2O\}_n$, FW = 524.83 g mol^{-1}: monoclinic, $C2/c$, a = 23.181(1) Å, b = 11. 3867(2) Å, c = 15.486(1) Å, β = 108.956(5)°, V = 3866.1(3) $Å^3$, Z = Z′ = 8,

ρ = 1.80 g cm^{-3}, $F(000)$ = 2112.0, R_{Bragg} = 0.047, R_p = 0.079 and R_{wp} = 0.110, for 4901 data and 53 parameters in the 7.0–105.0° 2θ range. CCDC No. 2038422.

4. Conclusions

In this work, we have described the synthesis and solid-state characterization of the novel coordination polymers (CPs) $[Zn(tr_2ad)Cl_2]_n$, $\{[Cu(tr_2ad)Cl]Cl{\cdot}4H_2O\}_n$, $[Cd_2(tr_2ad)Cl_4]_n$, $\{[Cu(tr_2ad)(NO_3)](NO_3)\}_n$ and $\{[Cd(tr_2ad)(NO_3)](NO_3){\cdot}H_2O\}_n$ [tr_2ad = 1,3-bis(1,2,4-triazol-4-yl)adamantane], isolated as air- and moisture-stable microcrystalline powders by means of solvothermal reactions. As assessed by thermogravimetric analysis, the five CPs show an appreciable thermal stability. As retrieved by powder X-ray diffraction, while $[Zn(tr_2ad)Cl_2]_n$ features 1-D chains, the other compounds contain 2-D double-layers. A comparative structural analysis involving known CPs built up with the tr_2ad ligand unveiled the coordination modes versatility of the ligand and the crystal structure dimensionality variability. Work can be anticipated in the functional characterization of these CPs as heterogeneous catalysts for cutting-edge organic reactions.

Supplementary Materials: The following are available online at http://www.mdpi.com/2304-6740/8/11/60/s1. Detailed synthesis of the tr_2ad ligand (Scheme S1). FTIR spectrum of tr_2ad (Figure S1). Ortep drawings for tr_2ad and $tr_2ad{\cdot}3H_2O$ (Figure S2). Further representation of the crystal structure of $\{[Cu(tr_2ad)Cl]Cl{\cdot}4H_2O\}_n$ (Figure S3). Comparison of bond distances at the metal ion (Figure S4). Graphical result of the final structure refinement carried out on $[Zn(tr_2ad)Cl_2]_n$, $\{[Cu(tr_2ad)Cl]Cl{\cdot}4H_2O\}_n$, $[Cd_2(tr_2ad)Cl_4]_n$, $\{[Cu(tr_2ad)(NO_3)](NO_3)\}_n$, and $\{[Cd(tr_2ad)(NO_3)](NO_3){\cdot}H_2O\}_n$ (Figures S5-S9). The CIF files of tr_2ad, $tr_2ad{\cdot}3H_2O$ and the five CPs, and the checkCIF output files of tr_2ad and $tr_2ad{\cdot}3H_2O$.

Author Contributions: Conceptualization, C.P. and S.G.; investigation: N.X., A.T., S.G., M.M., K.V.D. and G.A.S.; resources: K.V.D., C.P. and S.G.; writing—original draft preparation: A.T. and S.G.; writing—review and editing, A.T. and S.G.; supervision, C.P. and S.G. All authors have read and agreed to the published version of the manuscript.

Funding: This research received no external funding.

Acknowledgments: S.G. acknowledges Università dell'Insubria and C.P. acknowledges Università di Camerino for partial funding.

Conflicts of Interest: The authors declare no conflict of interest.

References

1. Hoskins, B.F.; Robson, R. Infinite polymeric frameworks consisting of three dimensionally linked rod-like segments. *J. Am. Chem. Soc.* **1989**, *111*, 5962–5964.
2. Morsali, A.; Hashemi, L. *Main Group Metal Coordination Polymers: Structures and Nanostructures*; John Wiley & Sons Inc.: Hoboken, NJ, USA, 2017.
3. Patel, V. *Synthesis, Characterization and Application of Coordination Polymers*; Lambert Academic Publishing: Saarbrücken, Germany, 2015.
4. Batten, S.R.; Neville, S.M.; Turner, D.R. *Coordination Polymers: Design, Analysis and Application*; Springer: New York, NY, USA, 2010.
5. Hong, M.-C.; Chen, L. (Eds.) *Design and Construction of Coordination Polymers*; John Wiley & Sons Inc.: Hoboken, NJ, USA, 2009.
6. Yaghi, O.M.; Kalmutzki, M.J.; Diercks, C.S. *Introduction to Reticular Chemistry: Metal-Organic Frameworks and Covalent Organic Frameworks*; Wiley-VCH: Weinheim, Germany, 2019.
7. Kaskel, S. (Ed.) *The Chemistry of Metal-Organic Frameworks: Synthesis, Characterization, and Applications*; Wiley-VCH: Weinheim, Germany, 2016.
8. MacGillivray, L.R.; Lukehart, C.M. (Eds.) *Metal-Organic Framework Materials*; Wiley-VCH: Weinheim, Germany, 2014.
9. Farrusseng, D. (Ed.) *Metal-Organic Frameworks: Applications from Catalysis to Gas Storage*; Wiley-VCH: New York, NY, USA, 2011.
10. Schröder, M. (Ed.) *Functional Metal-Organic Frameworks: Gas Storage, Separation and Catalysis*; Springer: Berlin, Germany, 2010.

11. Li, H.; Wang, K.; Sun, Y.; Lollar, C.T.; Li, J.; Zhou, H.-C. Recent advances in gas storage and separation using metal–organic frameworks. *Mater. Today* **2018**, *21*, 108–121.
12. Rogge, S.M.J.; Bavykina, A.; Hajek, J.; Garcia, H.; Olivos-Suarez, A.I.; Sepúlveda-Escribano, A.; Vimont, A.; Clet, G.; Bazin, P.; Kapteijn, F.; et al. Metal–organic and covalent organic frameworks as single-site catalysts. *Chem. Soc. Rev.* **2017**, *46*, 3134–3184. [PubMed]
13. Kuznetsova, A.; Matveevskaya, V.; Pavlov, D.; Yakunenkov, A.; Potapov, A. Coordination polymers based on highly emissive ligands: Synthesis and functional properties. *Materials* **2020**, *13*, 2699.
14. Heine, J.; Müller-Buschbaum, K. Engineering metal-based luminescence in coordination polymers and metal–organic frameworks. *Chem. Soc. Rev.* **2013**, *42*, 9232–9242. [PubMed]
15. Sun, L.; Campbell, M.G.; Dincă, M. Electrically conductive porous metal–organic frameworks. *Angew. Chem. Int. Ed.* **2016**, *55*, 3566–3579.
16. Mínguez Espallargas, G.; Coronado, E. Magnetic functionalities in MOFs: From the framework to the pore. *Chem. Soc. Rev.* **2018**, *47*, 533–557.
17. Liu, J.-Q.; Luo, Z.-D.; Pan, Y.; Singh, A.K.; Trivedi, M.; Kumar, A. Recent developments in luminescent coordination polymers: Designing strategies, sensing application and theoretical evidences. *Coord. Chem. Rev.* **2020**, *406*, 213145.
18. Zhang, Y.; Yuan, S.; Day, G.; Wang, X.; Yang, X.; Zhou, H.-C. Luminescent sensors based on metal-organic frameworks. *Coord. Chem. Rev.* **2018**, *354*, 28–45.
19. Zhang, X.; Wang, W.; Hu, Z.; Wang, G.; Uvdal, K. Coordination polymers for energy transfer: Preparations, properties, sensing applications, and perspectives. *Coord. Chem. Rev.* **2015**, *284*, 206–235.
20. Wu, M.-X.; Yang, Y.-W. Metal–organic framework (MOF)-based drug/cargo delivery and cancer therapy. *Adv. Mater.* **2017**, *29*, 1606134.
21. Janiak, C.; Vieth, J.K. MOFs, MILs and more: Concepts, properties and applications for porous coordination networks (PCNs). *New J. Chem.* **2010**, *34*, 2366–2388.
22. Robin, A.Y.; Fromm, K.M. Coordination polymer networks with O-and N-donors: What they are, why and how they are made. *Coord. Chem. Rev.* **2006**, *250*, 2127–2157.
23. Kitagawa, S.; Kitaura, R.; Noro, S.-I. Functional porous coordination polymers. *Angew. Chem. Int. Ed.* **2004**, *43*, 2334–2375.
24. Biradha, K.; Sarkar, M.; Rajput, L. Crystal engineering of coordination polymers using 4,4′-bipyridine as a bond between transition metal atoms. *Chem. Commun.* **2006**, 4169–4179. [CrossRef]
25. Gagnon, K.J.; Perry, H.P.; Clearfield, A. Conventional and unconventional metal–organic frameworks based on phosphonate ligands: MOFs and UMOFs. *Chem. Rev.* **2012**, *112*, 1034–1054.
26. Tabacaru, A.; Pettinari, C.; Galli, S. Coordination polymers and metal-organic frameworks built up with poly(tetrazolate) ligands. *Coord. Chem. Rev.* **2018**, *372*, 1–30.
27. Pettinari, C.; Tabacaru, A.; Galli, S. Coordination polymers and metal-organic frameworks based on poly(pyrazole)-containing ligands. *Coord. Chem. Rev.* **2016**, *307*, 1–31.
28. Liao, P.-Q.; He, C.-T.; Zhou, D.-D.; Zhang, J.-P.; Chen, X.-M. Porous Metal Azolate Frameworks. In *The Chemistry of Metal-Organic Frameworks: Synthesis, Characterization, and Applications*; Kaskel, S., Ed.; Wiley: Weinheim, Germany, 2016; Volume 1, pp. 309–343.
29. Liu, K.; Shi, W.; Cheng, P. The coordination chemistry of Zn(II), Cd(II) and Hg(II) complexes with 1,2,4-triazole derivatives. *Dalton Trans.* **2011**, *40*, 8475–8490.
30. Zhang, J.-P.; Zhang, Y.-B.; Lin, J.-B.; Chen, X.-M. Metal azolate frameworks: From crystal engineering to functional materials. *Chem. Rev.* **2012**, *112*, 1001–1033.
31. Pavlov, D.; Sukhikh, T.; Filatov, E.; Potapov, A. Facile synthesis of 3-(azol-1-yl)-1-adamantanecarboxylic acids—New bifunctional angle-shaped building blocks for coordination polymers. *Molecules* **2019**, *24*, 2717.
32. Aromí, G.; Barrios, L.A.; Roubeau, O.; Gamez, P. Triazoles and tetrazoles: Prime ligands to generate remarkable coordination materials. *Coord. Chem. Rev.* **2011**, *255*, 485–546.
33. Beckmann, U.; Brooker, S. Cobalt(II) complexes of pyridazine or triazole containing ligands: Spin-state control. *Coord. Chem. Rev.* **2003**, *245*, 17–29.
34. Klingele, M.H.; Brooker, S. The coordination chemistry of 4-substituted 3,5-di(2-pyridyl)-4*H*-1,2, 4-triazoles and related ligands. *Coord. Chem. Rev.* **2003**, *241*, 119–132.
35. Haasnoot, J.G. Mononuclear, oligonuclear and polynuclear metal coordination compounds with 1,2,4-triazole derivatives as ligands. *Coord. Chem. Rev.* **2000**, *200*, 131–185.

36. Billes, F.; Endrédi, H.; Keresztury, G. Vibrational spectroscopy of triazoles and tetrazole. *J. Mol. Struct. THEOCHEM* **2000**, *530*, 183–200.
37. Nakamoto, K. *Infrared and Raman Spectra of Inorganic and Coordination Compounds*, 6th ed.; Part B; John Wiley & Sons: Hoboken, NJ, USA, 2009.
38. Lever, A.B.P.; Mantovani, E.; Ramaswamy, B.S. Infrared combination frequencies in coordination complexes containing nitrate groups in various coordination environments. A probe for the metal–nitrate interaction. *Can. J. Chem.* **1971**, *49*, 1957–1964.
39. Nakamoto, K. *Infrared and Raman Spectra of Inorganic and Coordination Compounds*, 6th ed.; Part A; John Wiley & Sons: Hoboken, NJ, USA, 2009.
40. Senchyk, G.A.; Lysenko, A.B.; Babaryk, A.A.; Rusanov, E.B.; Krautscheid, H.; Neves, P.; Valente, A.A.; Gonçalves, I.S.; Krämer, K.W.; Liu, S.-X.; et al. Triazolyl–based copper–molybdate hybrids: From composition space diagram to magnetism and catalytic performance. *Inorg. Chem.* **2014**, *53*, 10112–10121.
41. Senchyk, G.A.; Lysenko, A.B.; Rusanov, E.B.; Chernega, A.N.; Jezierska, J.; Domasevitch, K.V.; Ozarowski, A. Structure and magnetic behavior of Cu^{II} MOFs supported by 1,2,4-triazolyl-bifunctionalized adamantane scaffold. *Eur. J. Inorg. Chem.* **2012**, *2012*, 5802–5813.
42. Boldog, I.; Lysenko, A.B.; Rusanov, E.B.; Chernega, A.N.; Domasevitch, K.V. 1,3,5-Triphenyladamantane and 1,3,5,7-tetraphenyladamantane. *Acta Crystallogr. C* **2009**, *65*, o248–o252.
43. Spek, A.L.J. Single-crystal structure validation with the program PLATON. *J. Appl. Crystallogr.* **2003**, *36*, 7–13.
44. Senchyk, G.A.; Lysenko, A.B.; Krautscheid, H.; Sieler, J.; Domasevitch, K.V. A dihydroxidotetracopper(II) framework supported by 4,4′-(adamantane-1,3-diyl)bis(1,2,4-triazole) and benzene-1,3,5-tricarboxylate bridges. *Acta Crystallogr. C* **2008**, *64*, m246–m249.
45. Senchyk, G.A.; Lysenko, A.B.; Krautscheid, H.; Rusanov, E.B.; Chernega, A.N.; Krämer, K.W.; Liu, S.-X.; Decurtins, S.; Domasevitch, K.V. Functionalized adamantane tectons used in the design of mixed-ligand copper(II) 1,2,4-triazolyl/carboxylate metal–organic frameworks. *Inorg. Chem.* **2013**, *52*, 863–872.
46. Senchyk, G.A.; Lysenko, A.B.; Rusanov, E.B.; Chernega, A.N.; Krautscheid, H.; Domasevitch, K.V. Polynuclear and polymeric metal complexes based upon 1,2,4-triazolyl functionalized adamantanes. *Inorg. Chim. Acta* **2009**, *362*, 4439–4448.
47. Senchyk, G.A.; Lysenko, A.B.; Naumov, D.Y.; Fedin, V.P.; Krautscheid, H.; Domasevitch, K.V. Multiple anion···π interactions with a soft selenium atom: Accommodation of $NCSe^-$ anions inside hydrophobic pockets of adamantane/1,2,4-triazole coordination framework. *Inorg. Chem. Commun.* **2010**, *13*, 1576–1579.
48. Senchyk, G.A.; Lysenko, A.B.; Boldog, I.; Rusanov, E.B.; Chernega, A.N.; Krautscheid, H.; Domasevitch, K.V. 1,2,4-Triazole functionalized adamantanes: A new library of polydentate tectons for designing structures of coordination polymers. *Dalton Trans.* **2012**, *41*, 8675–8689.
49. Lysenko, A.B.; Senchyk, G.A.; Lincke, J.; Lässig, D.; Fokin, A.A.; Butova, E.D.; Schreiner, P.R.; Krautscheid, H.; Domasevitch, K.V. Metal oxide-organic frameworks (MOOFs), a new series of coordination hybrids constructed from molybdenum(VI) oxide and bitopic 1,2,4-triazole linkers. *Dalton Trans.* **2010**, *39*, 4223–4231.
50. Lysenko, A.B.; Senchyk, G.A.; Domasevitch, K.V.; Hauser, J.; Fuhrmann, D.; Kobalz, M.; Krautscheid, H.; Neves, P.; Valente, A.A.; Goncalves, I.S. Synthesis and structural elucidation of triazolylmolybdenum(VI) oxide hybrids and their behavior as oxidation catalysts. *Inorg. Chem.* **2015**, *54*, 8327–8338.
51. Senchyk, G.A.; Lysenko, A.B.; Krautscheid, H.; Domasevitch, K.V. Fluoride molecular scissors: A rational construction of new Mo(VI) oxofluorido/1,2,4-triazole MOFs. *Inorg. Chem. Commun.* **2011**, *14*, 1365–1368.
52. Muñoz-Lara, F.J.; Gaspar, A.B.; Muñoz, M.C.; Lysenko, A.B.; Domasevitch, K.V.; Real, J.A. Fast detection of water and organic molecules by a change of color in an iron(II) microporous spin-crossover coordination polymer. *Inorg. Chem.* **2012**, *51*, 13078–13080.
53. Senchyk, G.A.; Bukhan'ko, V.O.; Lysenko, A.B.; Krautscheid, H.; Rusanov, E.B.; Chernega, A.N.; Karbowiak, M.; Domasevitch, K.V. Ag^I/V^V heterobimetallic frameworks generated from novel-type $\{Ag_2(VO_2F_2)_2(triazole)_4\}$ secondary building blocks: A new aspect in the design of SVOF hybrids. *Inorg. Chem.* **2012**, *51*, 8025–8033.
54. Sheldrick, G.M. A short history of SHELX. *Acta Crystallogr. A* **2008**, *64*, 112–122.
55. Sheldrick, G.M. Crystal structure refinement with SHELXL. *Acta Crystallogr. C* **2015**, *71*, 3–8.
56. Coelho, A.A. Indexing of powder diffraction patterns by iterative use of singular value decomposition. *J. Appl. Crystallogr.* **2003**, *36*, 86–95.
57. *Topas-R, V3*; Bruker AXS: Karlsruhe, Germany, 2005.

58. Coelho, A.A. Whole-profile structure solution from powder diffraction data using simulated annealing. *J. Appl. Crystallogr.* **2000**, *33*, 899–908.
59. Rietveld, H.M. A profile refinement method for nuclear and magnetic structures. *J. Appl. Crystallogr.* **1969**, *2*, 65–71.
60. Cheary, R.W.; Coelho, A.A. A fundamental parameters approach to X-ray line-profile fitting. *J. Appl. Crystallogr.* **1992**, *25*, 109–121.
61. Dollase, W.A. Correction of intensities for preferred orientation in powder diffractometry: Application of the March model. *J. Appl. Crystallogr.* **1986**, *19*, 267–272.

inorganics

MDPI

Article

Adsorption Properties of $Ce_5(BDC)_{7.5}(DMF)_4$ MOF

Cesare Atzori [1], Jayashree Ethiraj [2], Valentina Colombo [3], Francesca Bonino [1,*] and Silvia Bordiga [1]

1 Department of Chemistry, NIS and INSTM Reference Centre, Università di Torino, Via G. Quarello 15, I-10135 and Via P. Giuria 7, I-10125 Torino, Italy; cesare.atzori@unito.it (C.A.); silvia.bordiga@unito.it (S.B.)
2 National Centre for Catalysis Research and Department of Chemistry, Indian Institute of Technology Madras, Chennai 600036, India; jaijoy18@gmail.com
3 Department of Chemistry, Università degli Studi di Milano, Via C. Golgi 19, I-20133 Milano, Italy; valentina.colombo@unimi.it
* Correspondence: francesca.bonino@unito.it

Received: 7 November 2019; Accepted: 21 January 2020; Published: 26 January 2020

check for updates

Abstract: In this article we report on the spectroscopic and adsorptive studies done on Ce(III)-based MOF possessing, upon desolvation, open metal sites, and a discrete surface area. The Ce-based MOF was synthesized from terephthalic acid linker (H_2BDC) and Ce^{3+} cations by the classical solvothermal method. Preliminary powder X-ray diffraction analysis showed that the obtained materials corresponded to the ones reported by other authors. Spectroscopic techniques, such as XAS and in situ FTIR with probe molecules were used. In situ FTIR spectroscopy confirmed the successful removal of DMF molecules within the pore system at temperatures above 250 °C. Moreover, the use of CO as a probe molecule evidenced the presence of a Ce^{3+} open metal sites. Detailed volumetric and calorimetric CO_2 adsorption studies are also reported.

Keywords: cerium; MOF; terephthalic acid; spectroscopic characterization; adsorption; calorimetry; carbon dioxide

---

1. Introduction

Cerium is the most abundant lanthanide element present in the earth crust [1,2] and the ores that are mined for the extraction of more rare and precious rare earth elements (REEs) are also rich in Ce; thus, its cost is relatively low. Its oxide, CeO_2, commonly named ceria, is particularly relevant for redox chemistry, being a catalyst for oxidation and reduction reactions [3], for example, combustion catalysis [4] and photocatalysis [3].

Ce-based MOFs have recently created interest in the scientific community. General features that can be drawn from looking at the current published literature are the following: (i) both Ce^{3+} and Ce^{4+} oxidation states can be used in the synthesis of MOFs [5–9]; (ii) synthetic conditions for Ce^{3+}-containing MOFs tends to be harsher than Ce^{4+} [5,6,10]; (iii) usually, Ce^{4+} starting reagents may be reduced to Ce^{3+} during the synthesis [11,12]; (iv) Ce^{3+} materials more frequently have peculiar structures, while Ce^{4+} tends to give rise to MOFs with the same structure as other 4+ cations (e.g., Zr^{4+} or Hf^{4+}) [5–7,13,14]. Their thermal stability is generally lower than their Zr^{4+} counterparts [7,8,14,15]. As a possible application of Ce MOFs as redox catalysts, Smolders et al. [8] reported the successful use of Ce^{4+}-UiO-67 in the aerobic oxidation of benzylic alcohol to benzaldehyde mediated by TEMPO (2,2,6,6-Tetramethylpiperidinyloxyl). Furthermore, Ethiraj et al. [5] reported the use of a Ce^{3+}-based MOF for the selective capture and storage of CO_2, obtaining high figures of merit of capacity and separation.

The present work reports on the synthesis, spectroscopic characterization, and adsorption properties of Ce^{3+}-based MOFs with terephthalic acid (H_2BDC) as the linker. This material has been already introduced in the literature by D'Arras et al. [11], who discovered it and suggested its crystal structure in the as-synthesized form, together with the characterization of the thermal properties. However, the porosity of the material has not been studied, indeed, here we report on its adsorption properties, examined through spectroscopic techniques (XAS and in situ FTIR) and adsorption isotherms.

2. Results and Discussion

$Ce_5(BDC)_{7.5}(DMF)_4$ MOF [16] optimized synthesis is reported in the Supplementary Materials. The $Ce_5(BDC)_{7.5}(DMF)_4$ PXRD pattern was coincident with the one reported by D'Arras et al. [11]. The hypothesized structure was taken from Reference [11] and it is reported in Figure 1, for clarity. The asymmetric unit of crystalline structure shows chains of five independent cerium atoms arranged linearly and surrounded by BDC^{2-} and DMF molecules (see Figure 1a). The two terminal atoms of the group are coordinated by eight oxygen atoms in a distorted square antiprismatic shape: six oxygen atoms belong to BDC^{2-} and two oxygen atoms to DMF molecules, while the "central" three cerium atoms are coordinated with nine oxygens, all coming from the ligands, in an uncommon distorted shape. There are mono-dimensional or 1D channels formed in the structure parallel to the Ce chains along the crystallographic $1\bar{1}0$ direction. As depicted in Figure 1b, these micropores are mainly occupied by the coordinated DMF molecules, which protrude into the pores. The surface area and porosity of the material were clarified by the N_2 adsorption experiments (vide infra).

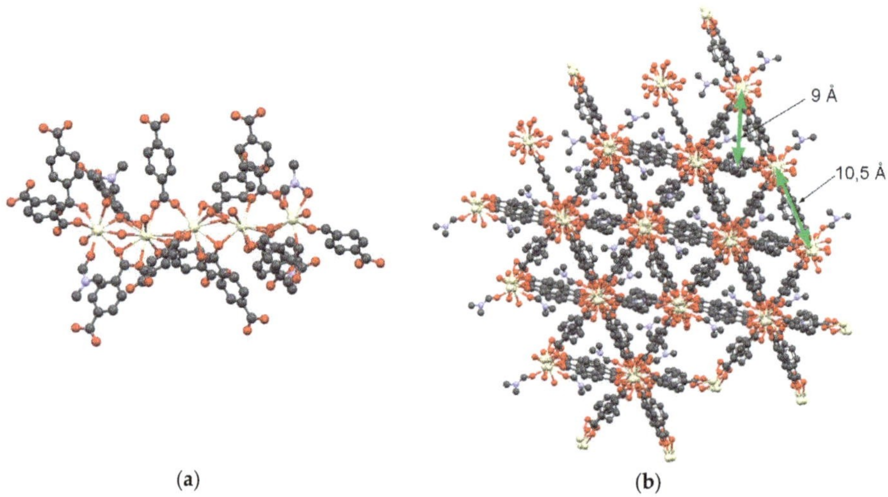

Figure 1. Structure of $Ce_5(BDC)_{7.5}(DMF)_4$. (**a**) Depiction of the asymmetric unit, containing five Ce atoms, 18 BDC^{2-}, and four DMF molecules. Cerium, carbon, nitrogen, and oxygen atoms are pale yellow, black, blue, and red respectively; hydrogen atoms are omitted for the sake of clarity; (**b**) a view through the $1\bar{1}0$ direction. Meaningful distances are highlighted.

The powder diffraction pattern of the as synthesized material, as shown in the synthesis development reported in the Supplementary Materials (Figures S1–S3), was compared to the calculated powder pattern obtained from the crystal structure reported by D'Arras et al. (Figure S3). SEM images of the synthesized powder are available in Figure S4.

A variable temperature powder X-ray diffraction (VTXRD) experiment in N_2 flow in the RT–600 °C temperature range (see Figure 2) showed that the material maintained the crystallinity until 475 °C, undergoing some changes in the XRD pattern, especially from 200 to 250 °C (highlighted in blue color), which could be due to the solvent removal, as strongly suggested by the TGA measurements reported in Figure S5.

The solution of the crystal structure of the desolvated material was out of the scope of the present work. At 500 °C the MOF started decomposing and at 525 °C the formation of broad diffraction peaks due to cerium dioxide was visible. The broadness of the peaks testifies that the particles were nanometric in size. Scherrer's equation [3], corrected by instrumental broadening using a Si standard from NIST, suggested a size for the cerium dioxide particles of 5 ± 1 nm.

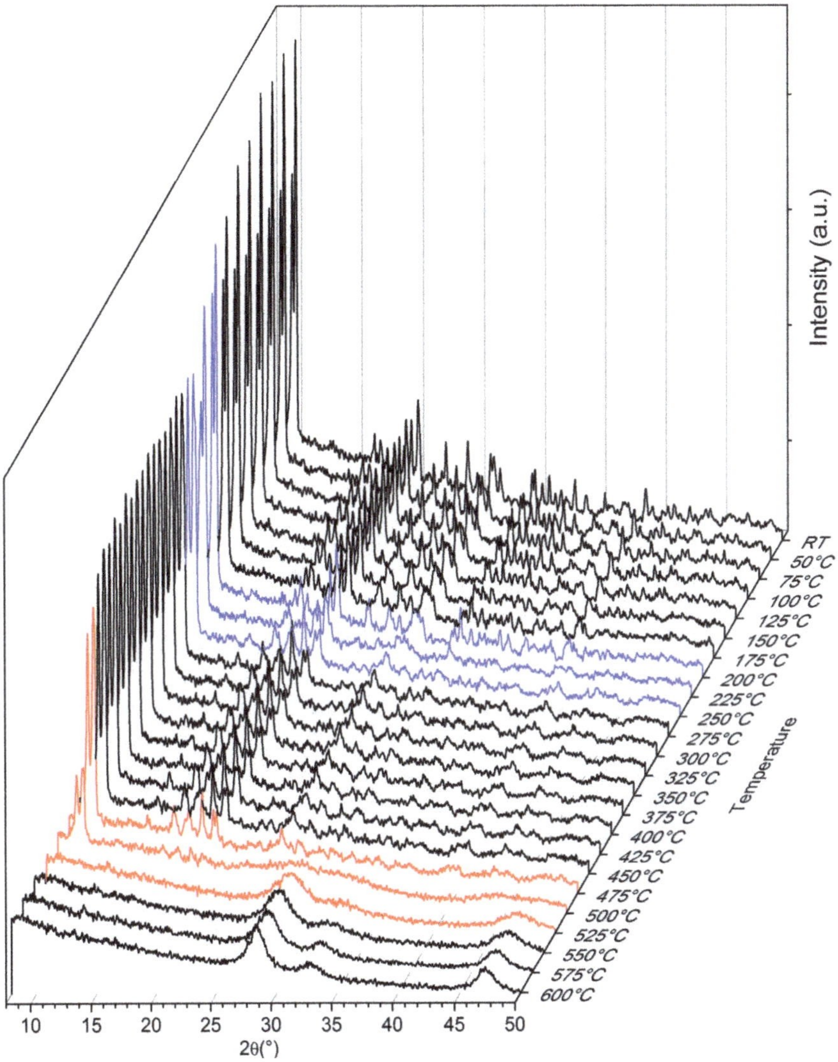

Figure 2. Variable temperature powder X-ray diffraction (VTXRD) recorded in the RT–600 °C range in N_2 flow.

The 3+ oxidation state of Ce in the as-synthesized state and activated at 350 °C material was confirmed by means of XAS spectroscopy, in the XANES region comparison with Ce^{3+} and Ce^{4+} standards. XANES spectra are reported in Figure 3. The 3+ oxidation state was also maintained after activation at 350 °C in the He stream directly in the measurement cell, in agreement with the results reported by the XPS experiments of D'Arras et al. [11].

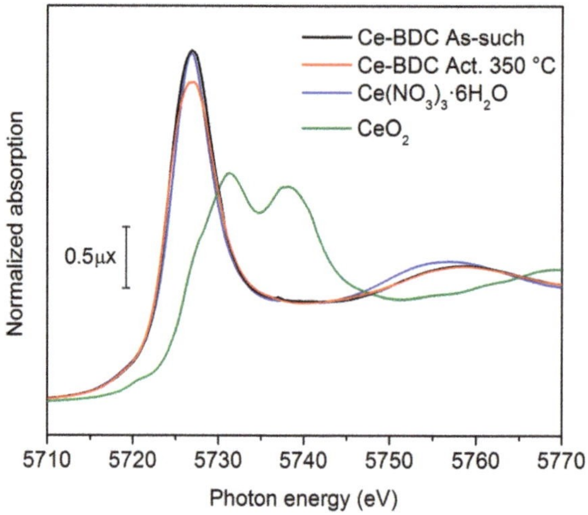

Figure 3. Ce L_3 edge XANES spectra of $Ce_5(BDC)_{7.5}(DMF)_4$ as synthesized and activated at 350 °C, compared with CeO_2 and $Ce(NO_3)_3{\cdot}6H_2O$.

N_2 adsorption volumetric isotherms were measured on the material in order to point out the specific surface area and porosity of the MOF. Thermal treatments were performed in vacuo for 3 h (a longer time in comparison with those ones performed in case of IR or XRD because of the bigger amount of sample) and were made in a consecutive way. From the isotherms reported in Figure 4 it is clear that upon the solvent loss (occurring in the 200–250 °C range) N_2 adsorption grew dramatically, showing microporosity (as the isotherm is a Type I) and higher surface area (more than 200 m^2/g) due to the accessible pores.

BET and Langmuir adsorption models for the surface area were applied; the results are summarized in Table 1. Generally, the reported value can be quite modest for MOF materials, compared with typical MOF surface areas (thousands of m^2/g) [17].

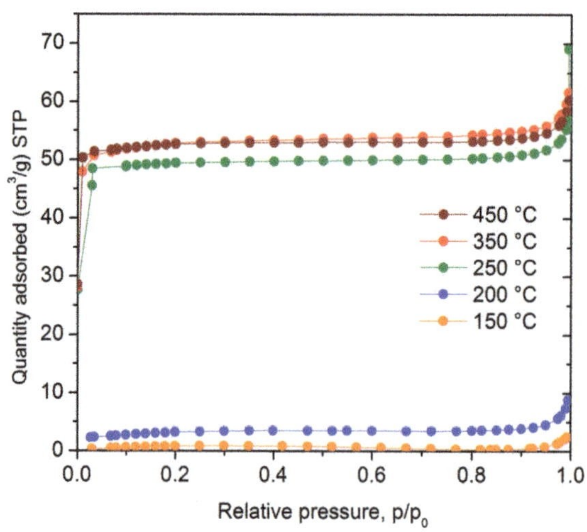

Figure 4. N_2 Adsorption isotherms at −196 °C measured on $Ce_5(BDC)_{7.5}(DMF)_4$. at different activation temperatures (150 °C, 200 °C, 250 °C, 350 °C, 450 °C).

Table 1. BET and Langmuir surface area for $Ce_5(BDC)_{7.5}(DMF)_4$ different temperature treatments.

Treatment	BET Surface Area (m^2/g)	C Value	Langmuir Surface Area (m^2/g)	t-Plot Micropore Volume (cm^3/g)
150 °C-3 h	3.63 ± 0.09	13.6	6.0 ± 0.2	-
200 °C-3 h	11.57 ± 0.06	115.9	16.1 ± 0.2	-
250 °C-3 h	223 ± 2	7110	217.3 ± 0.2	0.072
350 °C-3 h	212 ± 2	6893	232.9 ± 0.5	0.073
450 °C-3 h	216 ± 1	26,545	232 ± 0.4	0.075

As synthesized $Ce_5(BDC)_{7.5}(DMF)_4$ showed the typical mid-IR spectrum for a solvated MOF (see Figure 5). The typical vibrational fingerprints due to DMF solvent molecules mainly inside the pores and to H_2O molecules adsorbed from the atmosphere can be recognized: a broad band centered at 3400 cm^{-1} due to the hydrogen-bonded H_2O molecules and sharp features at frequencies lower than 3000 cm^{-1}, in the range of the aliphatic C–H stretching mode, and a very intense band centered at 1670 cm^{-1}, in the range of the carbonyl stretching mode, due to DMF molecules. Upon progressive outgassing, also by increasing the temperatures, vibrational signals due to DMF and H_2O disappeared and the typical spectrum of a carboxylate-based MOF material was shown: very intense bands in the 1650–1250 cm^{-1} range due to carboxylate stretching modes (both symmetrical and asymmetrical) and sharp features at frequencies higher than 3000 cm^{-1}, in the range of aromatic C–H stretching mode due to BDC^{2-}.

These data support the VTXRD, TGA (see Figure S5), and SSA experiments and therefore we can affirm the material is not destroyed even if activated at 450 °C, maintaining crystallinity and surface area, even if its crystal structure undergoes a phase transition that has not been determined in this study.

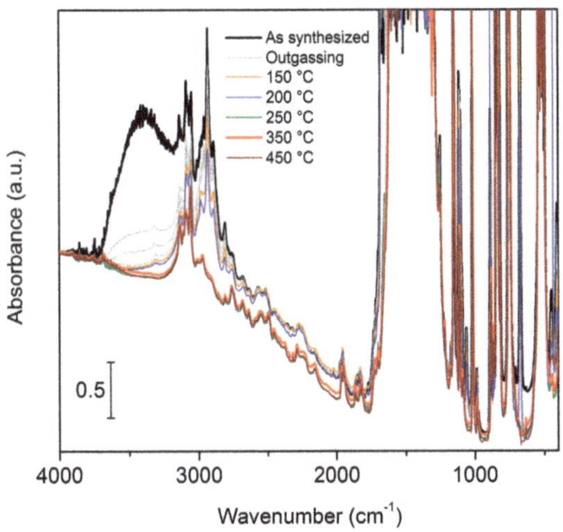

Figure 5. FTIR spectra of $Ce_5(BDC)_{7.5}(DMF)_4$ activated at different temperature in vacuo.

In situ CO adsorption FTIR spectra at −196 °C were recorded on a sample activated at 350 °C for 1 h (see Figure 6). A pressure of 5 mbar of CO was dosed (black spectrum). The lowest frequency peak, at 2131 cm^{-1}, was assigned to the physi-sorbed CO in the pores, as the first one to be desorbed. The other two bands at higher frequencies (2161 cm^{-1} and 2152 cm^{-1}) required more time for complete desorption, and for this reason they can be assigned to CO adsorbed on Lewis acidic sites.

Upon outgassing, the initial activated MOF spectrum was obtained due to the total reversibility of CO adsorption. Because the CO vibrational mode on the MOF open metal sites is intermediate between metals in oxides and metals grafted in different systems [18,19], the doublet can be assigned to CO adsorbed on Ce^{3+} sites, since CO interacting with Ce^{4+} is expected to give bands at frequencies higher than 2156 cm^{-1} [20,21], in agreement also with XANES results. With the crystal structure of the desolvated material unknown, we can only hypothesize the presence of at least two different Ce^{3+} probed sites [20–22]. Only two out of the five different cerium atoms among the crystallographic asymmetric unit underwent the removal of DMF molecules during activation, however for an in-depth understanding of the desolvation process more investigation is needed. It is of relevance also the overall low intensity of the bands associated with CO adsorption could be due to a low accessibility of the uncoordinated sites. In case of FTIR CO adsorption on MOF-76-Ce [5], two quite intense bands at 2155 cm^{-1} and 2149 cm^{-1}, due, respectively, to the probe interacting with one Ce site and bridged on two close sites, were reported, testifying the completely different local structure of the metal in MOF-76-Ce and in the present $Ce_5(BDC)_{7.5}(DMF)_4$.

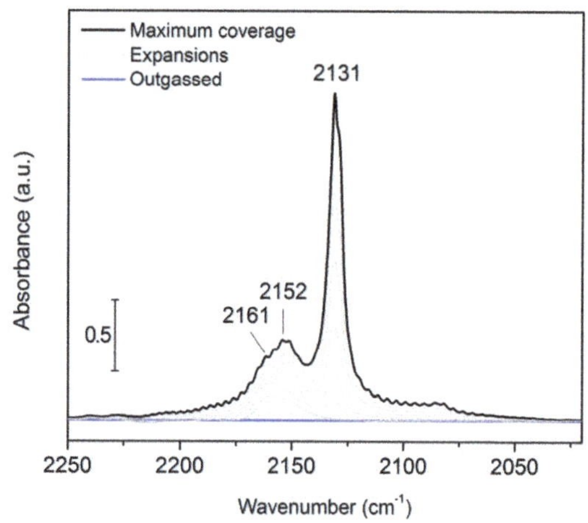

Figure 6. Background-subtracted CO adsorption FTIR spectra at −196 °C on $Ce_5(BDC)_{7.5}(DMF)_4$ activated at 350 °C for 1 h.

CO_2 volumetric isotherms at various activation temperatures are reported in Figure 7. The same powder was heated in vacuo for three hours at the next activation temperature in a consecutive way.

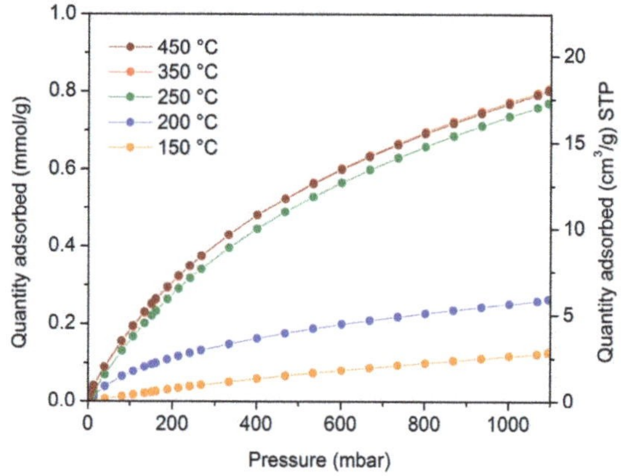

Figure 7. CO_2 Adsorption isotherms at 25 °C on $Ce_5(BDC)_{7.5}(DMF)_4$ at different activation temperatures (150 °C, 200 °C, 250 °C, 350 °C, 450 °C).

CO_2 uptake showed a clear increase with the activation temperature by reaching a plateau at the 350–450 °C temperature range, as reported in Figure 7 and Table 2.

Table 2. Summary of CO_2 uptake measurements at 1 bar and 25 °C.

Treatment	CO_2 Uptake at 1 Bar and 25 °C			$n_{ads}/S_{Langmuir}$(1 bar)
	(mol/kg)	(cm³/g) STP	Weight Percentage	(μmol/m²)
150 °C-3 h	0.1281	2.87	0.56%	21.4 ± 0.7
200 °C-3 h	0.2728	6.12	1.19%	16.93 ± 0.3
250 °C-3 h	0.7954	17.83	3.38%	3.660 ± 0.004
350 °C-3 h	0.8352	18.72	3.55%	3.586 ± 0.008
450 °C-3 h	0.8301	18.61	3.52%	3.586 ± 0.006

Calorimetric data, reported in Figure 8, were recorded for the adsorption of CO_2 at 30 °C in the 0–90 mbar range. The adsorbed quantity and the heat released by the adsorption as a function of the pressure are plotted, respectively, in Figure 8a,b. Both these curves are nearly Henry-type, as confirmed by the first part of our CO_2 volumetric adsorption isotherms collected with a different instrument at 25 °C (Figure 7). The studied pressure range was too low to observe saturation of the adsorbing sites and the temperature difference of 5 °C between these two experiments was due to technical requirements. The differential heat of adsorption at low coverages was about 32–33 kJ/mol and it is quite typical for the interaction of CO_2 with an open-metal site, as in MOF-76-ds [5], HKUST-1 [23], Mg-MOF-74 [17]. It is worth noting that the differential heat curve (Figure 8c) is characterized by an abrupt diminishing with the adsorbed quantity; this can be ascribed to the overall low number of strongly coordinating sites present in the desolvated material, as highlighted by our CO FTIR experiment (Figure 6). The total reversibility of the adsorption of CO_2 upon outgassing at 30 °C was testified thoroughly by the perfect recovery of the adsorption properties between the primary and secondary adsorption cycles.

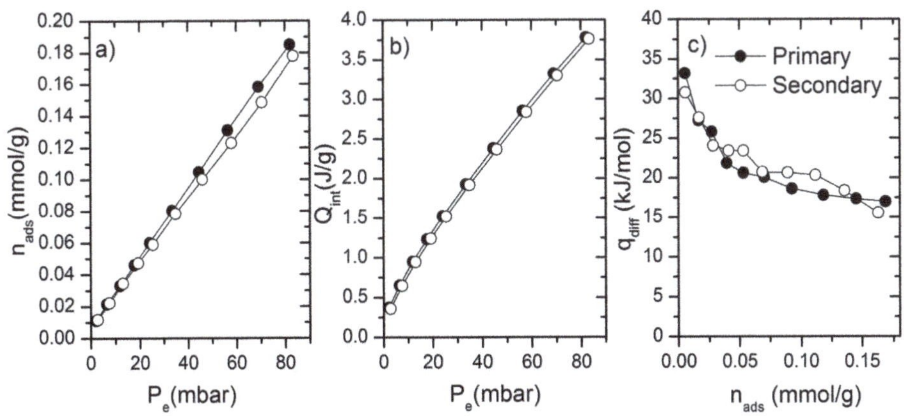

Figure 8. (**a**) Volumetric isotherm (**b**) calorimetric isotherm and (**c**) differential heat distribution of CO_2 adsorption measured at 30 °C on $Ce_5(BDC)_{7.5}(DMF)_4$ activated at 350 °C.

3. Materials and Methods

The variable temperature X-ray diffraction patterns (VTXRD) were collected with an X'Pert PRO MPD diffractometer from PANalytical (Almelo, The Netherlands), working in Bragg–Brentano geometry equipped with a Cu Kα source using about 10 mg of sample. Scattered photons were revealed by an X'celerator linear detector (PANalytical, Almelo, The Netherlands) equipped with a Ni filter to attenuate Kβ. A non-ambient chamber XRK900 from Anton Paar (Graz, Austria) with

Be windows was used to collect diffractograms as a function of temperature in a flow of dry N_2 (20 mL/min). The temperature program was set to measure a pattern every 25 °C, waiting 25 min at the target temperature before collecting the data. The temperature was increased at a rate of 2 °C/min.

X-Ray absorption spectra at the Ce L_3 edge (5723 eV) were collected at the BM23 beamline of the European Synchrotron Radiation Facility (ESRF). Data were acquired up to the Ce L_2 edge (6164 eV), which limited the EXAFS signal down to k ≈ 10 $Å^{-1}$.The acquisition step was set to 0.3 eV in the near-edge region and $\Delta k = 0.035$ $Å^{-1}$ in the EXAFS part of the spectrum. We used three He/N_2-filled ionization chambers as I_0, I_1, and I_2 detectors, placing chromium foil between I_1 and I_2 for energy calibration. XANES and EXAFS data were analyzed by using the Demeter 0.9.20 package. Sample treatment was carried out in an in situ cell under a flow of He (80 mL min^{-1}). During activation, the sample was heated to 350 °C after a ramp heating at 4 °C min^{-1}, to then be cooled to 30 °C.

Adsorption isotherms were collected on an ASAP 2020 apparatus from Micromeritics (Norcross, GA, USA) using a liquid nitrogen bath at −196 °C, albeit CO_2 isotherms were collected filling the same dewar vessel with water at 25 °C. About 150 mg of sample was heated in dynamic vacuum at 350 °C for 3 h prior to measuring the isotherm. Langmuir fit was made in the $0.05 < p/p_0 < 0.2$ range, while the BET analysis was carried out in a very low-pressure region, as prescribed by the so-called Rouquerol rules [24], in order to obtain a positive C value.

FTIR spectra were collected on a Nicolet 6700 from Thermo Scientific (Waltham, MA, USA) equipped with an MCT detector in the 4000–400 cm^{-1} range with a resolution of 2 cm^{-1}. The sample was prepared by pressing a thin self-supporting pellet (10 mg of sample) and using a vacuum line and a jacketed IR cell of local construction capable of cooling down the sample with liquid nitrogen and permitting the dosing of probe molecules (i.e., CO). The experiment was run by activating the pellet in dynamic vacuum at 350 °C for 1 h, then dosing about 15 mbar of CO on the sample to record spectra during cooling to the liquid nitrogen temperature, desorption, and then heating back to RT.

Adsorption heats were measured simultaneously with the adsorption isotherms by means of a C80 Tian-Calvet microcalorimeter from Setaram (Caluire-et-Cuire, France) at a temperature of 30 °C, coupled with a glass vacuum line of local construction. The procedure is thoroughly described in references [25,26], and in this case required thermal activation under dynamic vacuum at 350 °C for 3 h, then an overnight outgassing at 30 °C in the calorimeter before measuring the primary and the secondary adsorption runs in order to determine the non-desorbable (irreversible) fraction.

4. Conclusions

The synthesis of $Ce_5(BDC)_{7.5}(DMF)_4$ was successful, starting from $Ce(NO_3)_3{\cdot}6H_2O$ and H_2BDC in solvothermal conditions in DMF at 140 °C. We obtained the same crystal structure reported by D'Arras et al. [11] using a Ce^{3+} source directly, conversely to the previous contribution. The thermal stability (up to 475 °C in an inert atmosphere) previously observed by D'Arras et al. [11] was confirmed by our VTXRD and TGA measurements. Differently from D'Arras et al. [11] we found a discrete surface area due to microporosity through N_2 adsorption isotherms at −196 °C (about 220 m^2/g) after thermal activation in the 250–450 °C range. XAS confirmed the presence of Ce^{3+} in the material also after desolvation. FTIR spectroscopy confirmed the successful removal of DMF within the pore system at temperatures above 250 °C, and by means of low-temperature CO adsorption evidenced the presence of a Ce^{3+} open metal sites. The interaction of the desolvated material with CO_2 was characterized by volumetric and calorimetric measurements, finding a modest capacity of adsorption (about 3.5 wt % at 1 bar and 25 °C) but a relevant enthalpy (32–33 kJ/mol) for the very first dose, compatible with the presence of open metal sites. This work can open the way to a deep understanding and description of this MOF crystal structure and phase changes upon activation.

Supplementary Materials: The following are available online at http://www.mdpi.com/2304-6740/8/2/9/s1, Figure S1: Diffractograms of 1, 3, 4, and 6 batches: adopted solvent and metal to ratio (M:L) are reported, Figure S2: Diffractograms of 8, 12, 19 batches: reaction conditions are reported, Figure S3: Diffractograms of batch 19 and the MOF reported by D'Arras et al., Figure S4: SEM images of $Ce_5(BDC)_{7.5}(DMF)_4$ MOF. Part (b) reports a magnification of a portion reported in part (a), Figure S5: TGA in N_2 (solid line) and dry (dash-dotted line) air flow of $Ce_5(BDC)_{7.5}(DMF)_4$ MOF, Figure S6: Magnitude of the Fourier transform of k^2 $\chi(k)$ EXAFS signal (phase uncorrected) at different temperatures of $Ce_5(BDC)_{7.5}(DMF)_4$. $2.74 < k < 9.874$ Å^{-1} range for the transform is used, Figure S7: Background subtracted CO_2 adsorption FTIR spectra at RT on $Ce_5(BDC)_{7.5}(DMF)_4$ activated 350 °C for 1 h.

Author Contributions: Conceptualization, F.B. and J.E.; Methodology, F.B., J.E., and C.A.; Formal Analysis, C.A.; Investigation, C.A.; Resources, S.B.; Data Curation, C.A.; Writing—Original Draft Preparation, F.B. and C.A.; Writing—Review and Editing, all authors; Visualization, C.A. and F.B.; Validation, V.C.; Supervision, F.B. and S.B. All authors have read and agreed to the published version of the manuscript.

Funding: This research was funded by The Italian Ministry of Education, University and Research (MIUR), grant number 2017KKP5ZR (PRIN project). The APC was funded by the same institution.

Acknowledgments: Claudia Barolo is acknowledged for the fruitful discussion during the synthetic procedure optimization. The autors thank Jenny G. Vitillo for the help in collecting the adsorption data and Kirill A. Lomachenko for XAS data collection.

Conflicts of Interest: The authors declare no conflict of interest.

References

1. Haxel, G.B.; Hedrick, J.B.; Orris, G.J. *Rare Earth Elements—Critical Resources for High Technology Supporting Sound Management of Our Mineral Resources*; USGS: Reston, VA, USA, 2002.
2. Long, K.R.; Van Gosen, B.S.; Foley, N.K.; Cordier, D. *The Principal Rare Earth Elements Deposits of the United States: A Summary of Domestic Deposits and a Global Perspective*; USGS: Reston, VA, USA, 2012.
3. Montini, T.; Melchionna, M.; Monai, M.; Fornasiero, P. Fundamentals and Catalytic Applications of CeO_2-Based Materials. *Chem. Rev.* **2016**, *116*, 5987–6041. [CrossRef] [PubMed]
4. Kašpar, J.; Fornasiero, P.; Graziani, M. Use of CeO_2-based oxides in the three-way catalysis. *Catal. Today* **1999**, *50*, 285–298. [CrossRef]
5. Ethiraj, J.; Bonino, F.; Vitillo, J.G.; Lomachenko, K.A.; Lamberti, C.; Reinsch, H.; Lillerud, K.P.; Bordiga, S. Solvent-Driven Gate Opening in MOF-76-Ce: Effect on CO_2 Adsorption. *ChemSusChem* **2016**, *9*, 713–719. [CrossRef] [PubMed]
6. Atzori, C.; Lomachenko, K.A.; Øien-ØDegaard, S.; Lamberti, C.; Stock, N.; Barolo, C.; Bonino, F. Disclosing the Properties of a New Ce(III)-Based MOF: $Ce_2(NDC)_3(DMF)_2$. *Cryst. Growth Des.* **2019**, *19*, 787–796. [CrossRef]
7. Lammert, M.; Wharmby, M.T.; Smolders, S.; Bueken, B.; Lieb, A.; Lomachenko, K.A.; De Vos, D.; Stock, N. Cerium-based metal organic frameworks with UiO-66 architecture: Synthesis, properties and redox catalytic activity. *Chem. Commun.* **2015**, *51*, 12578–12581. [CrossRef]
8. Smolders, S.; Lomachenko, K.A.; Bueken, B.; Struyf, A.; Bugaev, A.L.; Atzori, C.; Stock, N.; Lamberti, C.; Roeffaers, M.B.J.; De Vos, D.E. Unravelling the Redox-catalytic Behavior of Ce^{4+} Metal–Organic Frameworks by X-ray Absorption Spectroscopy. *ChemPhysChem* **2018**, *19*, 373–378. [CrossRef]
9. Lammert, M.; Glißmann, C.; Reinsch, H.; Stock, N. Synthesis and Characterization of New Ce(IV)-MOFs Exhibiting Various Framework Topologies. *Cryst. Growth Des.* **2017**, *17*, 1125–1131. [CrossRef]
10. Griffin, S.L.; Wilson, C.; Forgan, R.S. Uncovering the structural diversity of Y(III) naphthalene-2,6-dicarboxylate MOFs through coordination modulation. *Front. Chem.* **2019**, *7*, 36. [CrossRef]
11. D'Arras, L.; Sassoye, C.; Rozes, L.; Sanchez, C.; Marrot, J.; Marre, S.; Aymonier, C. Fast and continuous processing of a new sub-micronic lanthanide-based metal–organic framework. *New J. Chem.* **2014**, *38*, 1477. [CrossRef]
12. Rhauderwiek, T.; Heidenreich, N.; Reinsch, H.; Øien-ØDegaard, S.; Lomachenko, K.A.; Rütt, U.; Soldatov, A.V.; Lillerud, K.P.; Stock, N. Co-Ligand Dependent Formation and Phase Transformation of Four Porphyrin-Based Cerium Metal–Organic Frameworks. *Cryst. Growth Des.* **2017**, *17*, 3462–3474. [CrossRef]
13. Lammert, M.; Reinsch, H.; Murray, C.A.; Wharmby, M.T.; Terraschke, H.; Stock, N. Synthesis and structure of Zr(IV)- and Ce(IV)-based CAU-24 with 1,2,4,5-tetrakis(4-carboxyphenyl)benzene. *Dalton Trans.* **2016**, *45*, 18822–18826. [CrossRef] [PubMed]
14. Lammert, M.; Glißmann, C.; Stock, N. Tuning the stability of bimetallic Ce(IV)/Zr(IV)-based MOFs with UiO-66 and MOF-808 structures. *Dalton Trans.* **2017**, *46*, 2425–2429. [CrossRef] [PubMed]

15. Lomachenko, K.A.; Jacobsen, J.; Bugaev, A.L.; Atzori, C.; Bonino, F.; Bordiga, S.; Stock, N.; Lamberti, C. Exact Stoichiometry of Ce_xZr_{6-x} Cornerstones in Mixed-Metal UiO-66 Metal–Organic Frameworks Revealed by Extended X-ray Absorption Fine Structure Spectroscopy. *J. Am. Chem. Soc.* **2018**, *140*, 17379–17383. [CrossRef] [PubMed]
16. Atzori, C. Synthesis and Characterization of a Ce-Based Metal–Organic Framework (MOF). Master's Thesis, Università di Torino, Turin, Italy, 2014.
17. Sumida, K.; Rogow, D.L.; Mason, J.A.; McDonald, T.M.; Bloch, E.D.; Herm, Z.R.; Bae, T.-H.; Long, J.R. Carbon Dioxide Capture in Metal–Organic Frameworks. *Chem. Rev.* **2012**, *112*, 724–781. [CrossRef]
18. Chavan, S.; Bonino, F.; Vitillo, J.G.; Groppo, E.; Lamberti, C.; Dietzel, P.D.C.; Zecchina, A.; Bordiga, S. Response of CPO-27-Ni towards CO, N_2 and C_2H_4. *Phys. Chem. Chem. Phys.* **2009**, *11*, 9811–9822. [CrossRef]
19. Bordiga, S.; Regli, L.; Bonino, F.; Groppo, E.; Lamberti, C.; Xiao, B.; Wheatley, P.S.; Morris, R.E.; Zecchina, A. Adsorption properties of HKUST-1 toward hydrogen and other small molecules monitored by IR. *Phys. Chem. Chem. Phys.* **2007**, *9*, 2676–2685. [CrossRef]
20. Vindigni, F.; Manzoli, M.; Tabakova, T.; Idakiev, V.; Boccuzzi, F.; Chiorino, A. Effect of ceria structural properties on the catalytic activity of Au–CeO_2 catalysts for WGS reaction. *Phys. Chem. Chem. Phys.* **2013**, *15*, 13400. [CrossRef]
21. Binet, C.; Daturi, M.; Lavalley, J.C. IR study of polycrystalline ceria properties in oxidised and reduced states. *Catal. Today* **1999**, *50*, 207–225. [CrossRef]
22. Chavan, S.M.; Shearer, G.C.; Svelle, S.; Olsbye, U.; Bonino, F.; Ethiraj, J.; Lillerud, K.P.; Bordiga, S. Synthesis and Characterization of Amine-Functionalized Mixed-Ligand Metal–Organic Frameworks of UiO-66 Topology. *Inorg. Chem.* **2014**, *53*, 9509–9515. [CrossRef]
23. Grajciar, L.; Bludský, O.; Nachtigall, P. Water adsorption on coordinatively unsaturated sites in CuBTC MOF. *J. Phys. Chem. Lett.* **2010**, *1*, 3354–3359. [CrossRef]
24. Thommes, M.; Kaneko, K.; Neimark, A.V.; Olivier, J.P.; Rodriguez-Reinoso, F.; Rouquerol, J.; Sing, K.S.W. Physisorption of gases, with special reference to the evaluation of surface area and pore size distribution (IUPAC Technical Report). *Pure Appl. Chem.* **2015**, *87*, 1051–1069. [CrossRef]
25. Auroux, A. *Calorimetry and Thermal Methods in Catalysis*; Auroux, A., Ed.; Springer Series in Materials Science; Springer: Berlin/Heidelberg, Germany, 2013; Volume 154, ISBN 978-3-642-11953-8.
26. Bolis, V.; Maggiorini, S.; Meda, L.; D'Acapito, F.; Turnes Palomino, G.; Bordiga, S.; Lamberti, C. X-ray photoelectron spectroscopy and X-ray absorption near edge structure study of copper sites hosted at the internal surface of ZSM-5 zeolite: A comparison with quantitative and energetic data on the CO and NH_3 adsorption. *J. Chem. Phys.* **2000**, *113*, 9248–9261. [CrossRef]

inorganics

MDPI

Article

An Optimised Compaction Process for Zr-Fumarate (MOF-801)

Marco Taddei *, Matthew J. McPherson, Abel Gougsa, Jamie Lam, Jack Sewell and Enrico Andreoli

Energy Safety Research Institute, Swansea University, Fabian Way, Swansea SA1 8EN, UK
* Correspondence: marco.taddei@swansea.ac.uk

Received: 29 July 2019; Accepted: 31 August 2019; Published: 5 September 2019

Abstract: We reported a systematic approach aimed at identifying the optimal conditions for compaction of MOF-801, a small-pore zirconium-based metal–organic framework (MOF) containing fumaric acid as the linker, that can be easily synthesised in aqueous medium. Pellets of the MOF were prepared by compressing the powder either in neat form or dry-mixed with binders (sucrose, polyvinylalcohol, polyvinylbutyral) under a range of pressures and for different times. The mechanical stability and durability of the pellets was tested by simple drop tests and shake tests, finding that addition of 5% of polyvinylbutyral was enough to produce highly resilient pellets that did not release significant amounts of powder upon cracking. The crystallinity, textural properties and CO_2 adsorption performance of the MOF were successively assessed, observing the least change of the original properties in pellets compressed at 146 MPa for 15 s. Compaction at higher pressures impacted the performance more heavily, with no evident benefit from the mechanical point of view, whereas compression time did not have a relevant effect. The cyclic adsorption behaviour was tested, showing that the pellets retained as much as 90% of the CO_2 working capacity, while displaying unaffected sorption kinetics, and 74% of the H_2O working capacity.

Keywords: metal–organic frameworks; solid sorbents; shaping; gas separation; gas storage; water harvesting; zirconium

---

1. Introduction

Since their discovery over two decades ago, metal–organic frameworks (MOFs) have been considered promising materials for application in gas separation and storage, owing to their large porosity and ease of functionalisation [1–3]. Deployment of MOFs as solid sorbents for application in industrial-scale gas separation or storage processes requires powders to be compacted into industrially acceptable forms, such as tablets, pellets or extrudates. The use of free-flowing fine powders in industrial scale gas separation processes is indeed not desirable, mainly due to the large pressure drop across the sorbent bed [4,5]. In storage applications, high volumetric adsorption capacity is a key requirement to minimize the size of the gas container, calling for densification of the sorbent [6]. In addition to these technological aspects, there are several safety concerns arising from handling of free-flowing powders [5]. However, compaction of MOF powders under pressure often leads to partial collapse of their porous framework, thus reducing surface area and pore volume, with an impact on the gas sorption performance [7–9]. The challenge is therefore to identify the correct conditions to ensure mechanical stability of the shaped sorbents while preserving the performance of the original powder. Given its practical importance, there is a growing interest in the shaping of powders in the MOF community.

Zirconium-based MOFs are one of the most intensively investigated classes of MOFs, owing to their exceptional stability and topological diversity [10]. MOF-801, based on fumaric acid as the organic linker, features twelve-connected $[Zr_6O_4(OH)_4]^{12+}$ clusters, giving rise to **fcu** topology (Figure 1) [11–13]. Thanks to the high solubility of fumaric acid in water, MOF-801 can conveniently be synthesised in aqueous medium, with obvious benefits in terms of environmental impact [12]. This aspect, combined with the commercial availability and very low cost of fumaric acid, makes MOF-801 a strong candidate for large scale production. In terms of applications, MOF-801 is widely recognised as one of the best MOFs for adsorbing water directly from the air [13,14], besides displaying good CO_2 uptake capacity [15,16]. To the best of our knowledge, there is no report of a systematic approach aimed at identifying the optimal conditions for compaction of MOF-801. Therefore, we set out to develop a compaction process potentially transferable to a real industrial setting and able to afford durable pellets retaining as much as possible of the original gas sorption performance of the MOF in powder form.

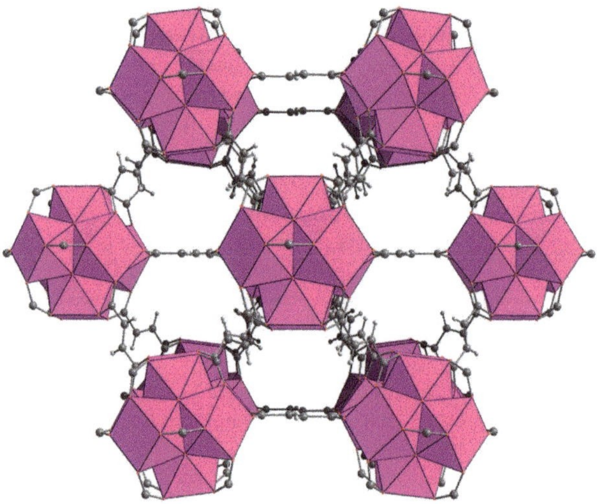

Figure 1. Crystal structure of MOF-801. Colour code: Zr, pink; C, grey; O, red; H, white.

2. Results and Discussion

To start our investigation, we employed the conditions recently reported by Bambalaza et al. [17] to produce pellets of UiO-66, which is closely related to MOF-801, having the same topology but a different linker (terephthalic acid). We prepared four MOF-801 pellets by compressing 400 mg of MOF for 15 min under four different pressures (146, 292, 438 and 584 MPa) using a Retsch PP25 pellet press with a 16-mm die. The pellets were 1.3-mm thick, which led to the calculation of an aspect ratio of 12.3. Powder X-ray diffraction (PXRD) analysis of the pellets (Figure 2) showed that the reflections of MOF-801 underwent progressive broadening as the pressure increased, suggesting partial loss of long-range order. In addition, a diffuse scattering feature below 10° 2θ appeared in the pellets, suggesting that part of the solid might have undergone more severe amorphisation. Interestingly, we observed that, upon compression, the MOF released water which was originally trapped in its pores. Therefore, we treated 400 mg of MOF-801 at 120 °C for 2 h to remove most of the water present in the structure and prepared a pellet compressing the warm powder at 438 MPa for 15 min, observing a much larger drop in the intensity of MOF-801 reflections (Figure S1). This suggests that the water present within the pores might act as a "cushion" and help to evenly dissipate the applied pressure, preventing the collapse of the framework during compression. Similar behaviour was previously reported for MOF-5 [18] and HKUST-1 [19,20]. The durability of the pellets prepared using the solvated MOF was assessed with simple drop tests. The pellets were dropped from a height of 36 cm until they underwent severe damage, intended here as breakage into more than two fragments. This choice was justified by the fact that the shape of the pellets is highly anisotropic and the meridian is by far the weakest direction along which breakage would occur. Overall, the pellets were quite brittle, failing after five drops at most and cracking into several tiny fragments (Figure S2). Increasing the pressure from 146 to 584 MPa yielded a little improvement in mechanical strength.

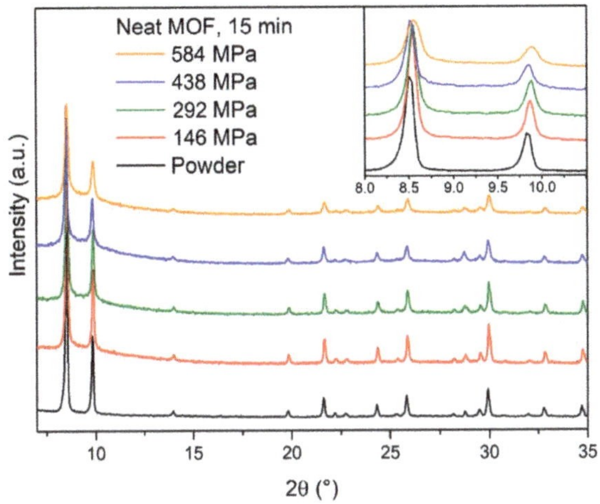

Figure 2. Powder X-ray diffraction (PXRD) patterns of MOF-801 powder (black) and pellets prepared compressing the same powder for 15 min under 146 (red), 292 (olive), 438 (blue) and 584 (orange) MPa pressure. The inset shows the low angle region.

The addition of binders was then considered, initially adding 5 wt % of either polyvinylalcohol (PVA), polyvinylbutyral (PVB) or sucrose, whose structures are shown in Figure 3, compressing at 438 MPa for 15 min. The choice of these binders was based on recent literature reports where they were employed to produce pellets of MOFs [21–23]. Before characterisation, the pellets containing binders were treated at 120 °C for 2 h. This treatment was applied under the assumption that in a real gas separation/storage application, the pellets would need to be activated at high temperature to make their porosity accessible. Figure 4 shows photographs of representative pellets prepared either with neat MOF-801 or with the addition of binders. All the pellets containing binders underwent colour change upon thermal treatment, with the ones containing PVA and sucrose showing the most evident effect, suggesting that these two binders might suffer from thermal degradation.

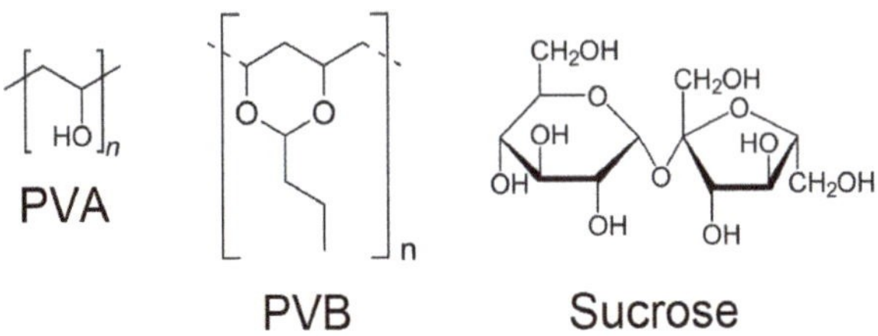

Figure 3. Molecular formulas of the binders employed in this study. From left to right: polyvinylalcohol (PVA), polyvinylbutyral (PVB) and sucrose.

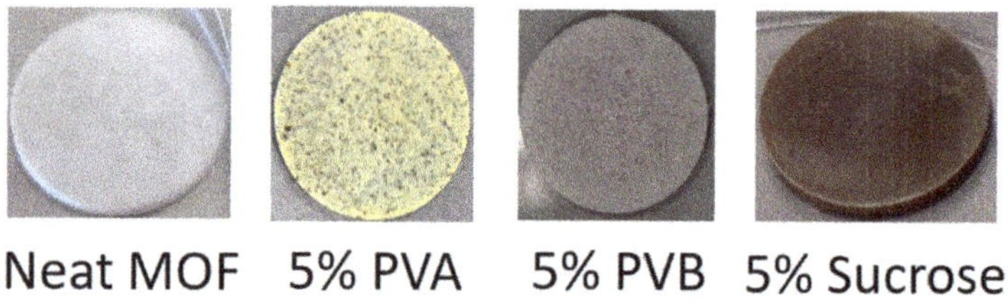

Figure 4. Photographs of representative pellets after thermal treatment at 120 °C for 2 h.

PXRD analysis does not evidence any relevant difference between the neat and the binder-containing pellets (Figure 5). In terms of durability, sucrose did not have any benefit and the pellets were as brittle as those made of neat MOF, whereas PVA and PVB had a clear strengthening effect and the pellets cracked into larger fragments, with no release of fine powder (Figure 6, Figures S3–S5). Scanning electron microscopy (SEM) images of the section of cracked pellets, obtained by compressing MOF-801 at 438 MPa for 2 min either neat or blended with 5% PVB, were acquired using backscattered electrons to enhance the contrast between the MOF and the binder. The neat pellet appears homogeneously compacted and relatively smooth (Figure 7a,c), whereas the one containing the binder displayed a rougher texture and dark spots of binder with size <10 μm dispersed throughout the MOF matrix (Figure 7b,d). Since PVB displayed the highest durability, we decided to carry out additional screening focusing on this binder.

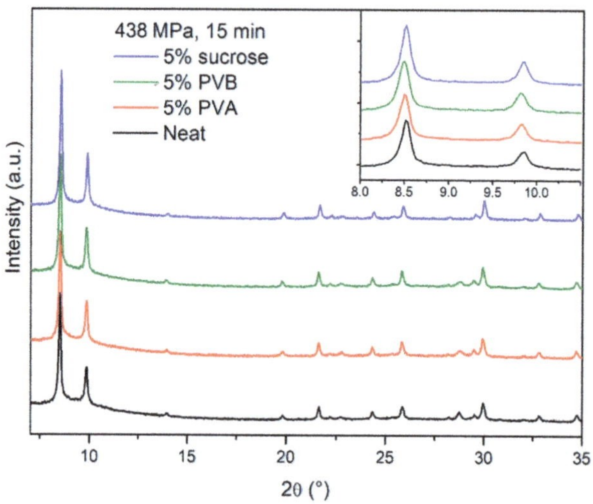

Figure 5. PXRD patterns of pellets prepared compressing neat MOF-801 (black), MOF-801 blended with 5% PVA (red), MOF-801 blended with 5% PVB (olive) and MOF-801 blended with 5% sucrose (blue) for 15 min under 438 MPa pressure. The inset shows the low angle region.

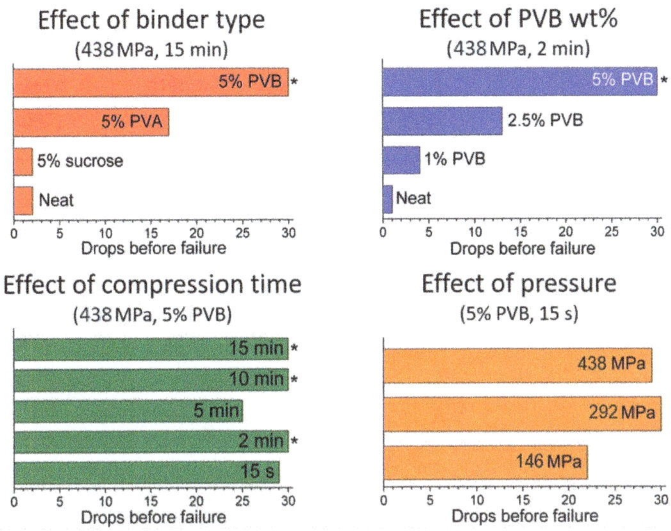

Figure 6. Results of drop tests: effect of binder type (top left), effect of binder amount (top right), effect of compression time (bottom left) and effect of pressure (bottom right). The symbol * means that no failure occurred.

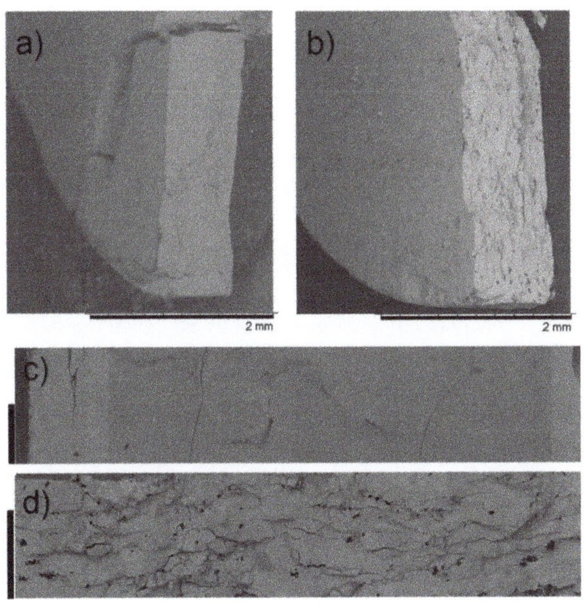

Figure 7. SEM micrographs of pellets prepared compressing at 438 MPa for 2 min using neat MOF-801 (**a**,**c**) and a mixture containing 5 wt % PVB (**b**,**d**). For (**c**,**d**), the scale bars on the left-hand side are 300 μm.

We then moved on to investigate the effect of compression time, with the aim of reducing this to a minimum. This is indeed important in a real compaction process, e.g., using a rotary pellet press, where it is not practical to keep the powder under pressure for as long as 15 min because of the low throughput of pellets produced in such conditions. We progressively decreased the compression time down to 10 min, 5 min, 2 min and 15 s, while keeping pressure to 438 MPa, observing no evident difference in the PXRD patterns (Figure 8). Drop tests for these pellets did not show any evidence of obvious dependence of mechanical strength on the compression time, suggesting that 15 s was a viable compression time in terms of durability, while enabling a much higher throughput than the initial 15 min (Figure 6). Using a hand-operated pellet press does not allow the further decrease of compression time in an accurate manner, but we speculate that time could be further decreased in a real, continuous compaction process without compromising the durability of the pellets.

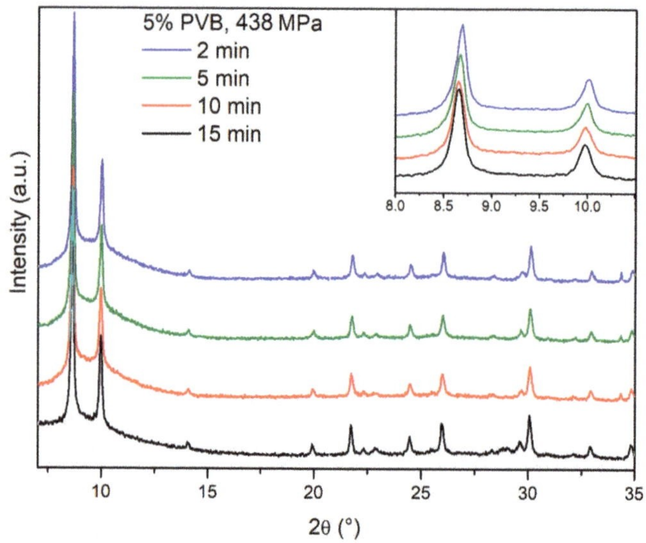

Figure 8. PXRD patterns of pellets prepared compressing MOF-801 blended with 5% PVB under 438 MPa of pressure for 15 min (black), 10 min (red), 5 min (olive) and 2 min (blue). The inset shows the low angle region.

The next parameter we screened was the amount of binder added, preparing pellets containing 2.5 and 1% of PVB and compressing at 438 MPa for 2 min. Drop tests showed that the durability scales almost linearly with the amount of binder (Figure 6), suggesting that our initial choice of using 5% PVB was a good compromise, able to provide durable pellets while keeping the binder loading to a relatively low value, thus limiting its effect on the adsorption performance of the pellets.

Finally, we decreased the pressure while keeping the compression time fixed at 15 s. This is an important parameter in a production process, since the materials constituting the pellet press must be able to withstand the repeated stress induced during compression over a long period of time. Reducing the pressure needed to obtain pellets is therefore crucial in limiting the equipment capital cost. We observed that the pellets compressed at 146 MPa displayed lower mechanical strength than those at 292 and 438 MPa, but their durability was still reasonable (more than 20 drops were needed for severe damage, with no release of fine powder). In order to assess the effect of aspect ratio on the durability of pellets, a 5-mm diameter die was employed, compressing 70 mg of MOF blended with 5% PVB at 146 MPa for 15 s. Pellets with a thickness of 2 mm and an aspect ratio of 2.5 were obtained. These pellets proved to be extremely tough, showing no signs of damage after 30 drops. In addition, we tested the resistance of our 16 mm pellets to attrition by loading them in a capped glass vial and using a vortex mixer to shake it for 30 s up to four times. After each cycle, the pellets were weighed to determine how much mass had been lost to fine powder, observing a less than 4% decrease for the PVB-containing pellets compressed at 146 and 263 MPa (Table 1).

Table 1. Results of attrition tests for 16-mm diameter pellets.

Pellet	Cumulative Weight Loss			
	Cycle 1	Cycle 2	Cycle 3	Cycle 4
Neat, 438 MPa, 2 min	6.4%	8.8%	9.3%	9.4%
5% PVB, 438 MPa, 15 s	2.4%	4.6%	6.3%	7.7%
5% PVB, 292 MPa, 15 s	1.1%	2.3%	3.0%	3.3%
5% PVB, 146 MPa, 15 s	2.3%	3.8%	3.8%	3.8%

The three 16-mm diameter pellets prepared using 5% PVB compressing for 15 s at 146, 292 and 438 MPa, respectively, were characterised for their gas sorption properties and compared with the original MOF-801 powder. N_2 sorption isotherms at 77 K, shown in Figure 9, display a decrease in uptake when the powder was compressed into pellets, with no major differences between different pressures. The shape of the isotherms suggests that the MOF retained its microporosity when compressed and that compaction reduced interparticular porosity (as indicated by the reduction of condensation occurring at high relative pressure). The BET surface area of the MOF-801 powder was 899 $m^2\ g^{-1}$, whereas the pellets displayed 569 $m^2\ g^{-1}$ (146 MPa), 546 $m^2\ g^{-1}$ (292 MPa) and 605 $m^2\ g^{-1}$ (438 MPa).

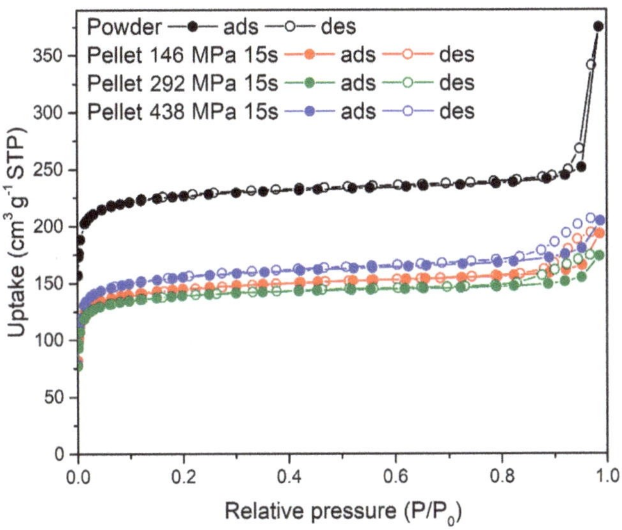

Figure 9. N_2 sorption isotherms at 77 K for MOF-801 powder (black) and pellets containing 5 wt % PVB prepared by compressing for 15 s under 146 (red), 292 (olive) and 438 (blue) MPa pressure.

The CO_2 uptake capacity in dynamic conditions was also evaluated, exposing both the powder and the pellets to 1 atm CO_2 at 40 °C in a thermogravimetric analyzer (Figure 10). The samples were previously activated by heating at 120 °C for 2 h under an Ar atmosphere, followed by cooling to 40 °C and switch to a CO_2 flow. The powder reached an uptake of 6.70 wt % after 35 min exposure, whereas the pellets reached 5.74 wt % (corresponding to 86% of the powder uptake; 146 MPa), 4.67 wt % (corresponding to 70% of the powder uptake; 292 MPa) and 4.47 wt % (corresponding to 67% of the powder uptake; 438 MPa), respectively. The adsorption kinetics were not significantly influenced upon compression of the powder, with all the samples reaching 50% and 90% of their maximum uptake within 6.8–7.3 min and 21.1–21.8 min, respectively. The stability over ten adsorption-desorption cycles was also tested for the powder and the pellet compressed at 146 MPa (Figure 11). The activated samples were exposed to 1 atm CO_2, swinging the temperature between 40 °C and 90 °C. In these conditions, both samples displayed consistent working capacity over ten cycles: 4.64 wt % for the powder and 4.18 wt % for the pellet (corresponding to 90% of the working capacity of the powder). We note that the baseline gain after the first adsorption cycle is due to the fact that, during the first adsorption cycle, the MOF was completely desolvated and able to adsorb about 6 wt % of CO_2 at 40 °C. Heating the MOF to 90 °C under an atmosphere of CO_2 did not completely remove all the adsorbed CO_2, leading to the working capacities reported above.

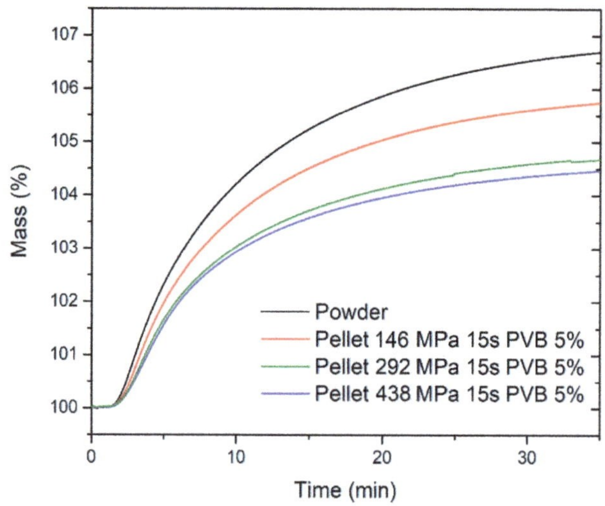

Figure 10. Dynamic CO_2 adsorption curves for MOF-801 powder (black) and pellets containing 5 wt % PVB prepared by compressing for 15 s under 146 (red), 292 (olive) and 438 (blue) MPa pressure. The experiments were performed under 1 atm CO_2. The delayed onset of adsorption was due to non-immediate saturation of the TGA chamber with CO_2 upon switching from Ar after activation.

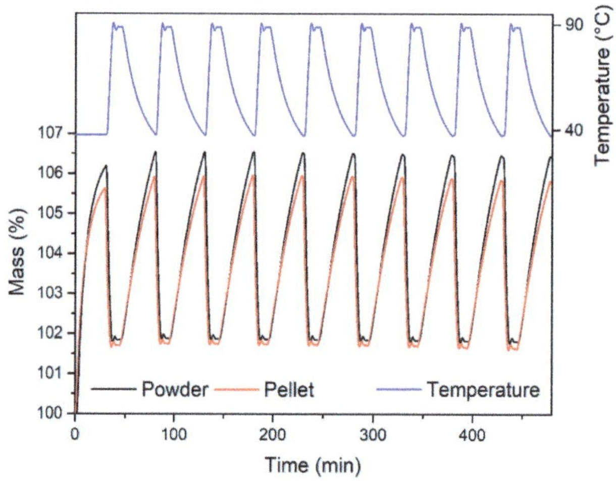

Figure 11. Cyclic CO_2 uptake experiments for MOF-801 free-flowing powder (black) and the pellet containing 5 wt % PVB prepared by compressing for 15 s under 146 MPa pressure (red). The entirety of the experiments were performed under 1 atm CO_2.

Finally, we compared the cyclic H_2O adsorption behaviour, exposing the powder and the pellet compressed at 146 MPa to a stream of humid air at 40 °C, regenerating by heating under Ar at 120 °C (Figure 12). The powder displayed a constant working capacity of 35.5% over ten cycles. The pellet displayed a lower initial working capacity of 26.4% (corresponding to 74% of the working capacity of the powder), which slightly reduced to 24.1% over ten cycles.

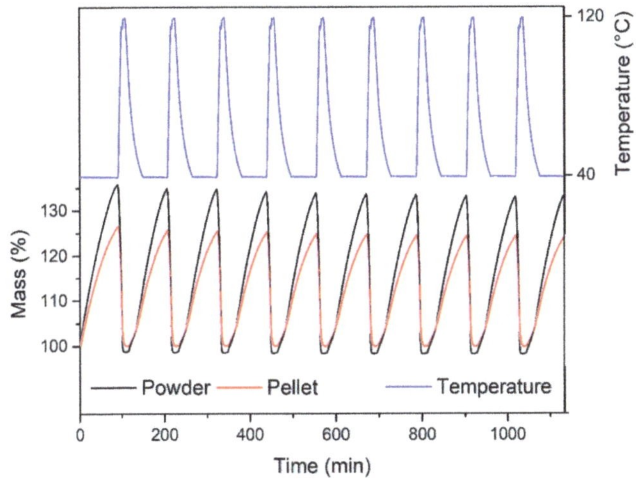

Figure 12. Cyclic H_2O adsorption experiments for MOF-801 free-flowing powder (black) and the pellet containing 5 wt % PVB prepared by compressing for 15 s under 146 MPa pressure (red). The experiments were performed under 1 atm humidified air in adsorption and 1 atm Ar in desorption.

3. Materials and Methods

3.1. Chemicals

Zirconium(IV) chloride (98%, anhydrous, Acros Organics, Geel, Belgium), formic acid (≥98%, Sigma Aldrich, Darmstadt, Germany), fumaric acid (≥99.0%, Sigma Aldrich), Polyvinyl butyral (Sigma-Aldrich), Poly(vinyl alcohol) (Mw 31,000–50,000, 98–99% hydrolysed, Sigma Aldrich), Sucrose (Sigma-Aldrich, ≥99.5%).

3.2. Synthesis of MOF-801

In a 500-mL round-bottom flask, water (377 mL) and formic acid (78 mL, 2.1 mol) were combined with stirring. To this mixture, zirconium chloride (4.8 g, 2.1 mmol) was added in portions and the solution stirred until clear. Fumaric acid (7.2 g, 6.2 mmol) was then added and the solution stirred until clear. The mixture was then heated to reflux and allowed to reflux for 72 h. Then, the solution was cooled and the solid recovered by dividing the reaction mixture into centrifuge tubes and centrifuging (5000 rpm, 5 min) until supernatant solution was clear. The solid was then redispersed in 50:50 water:ethanol mixture (50 mL per tube) while being agitated with a tilt table for 1 h. This process was repeated once and then the solid was dispersed in acetone (50 mL per tube) and agitated in the same manner for 1 h, repeating once. The solid was then allowed to dry at 65 °C overnight. Yield: 4.0 g.

3.3. Pellet Preparation

A Retsch PP25 pellet press was employed for compaction of MOF-801 powders. For pellets of 16 mm diameter, either 400 mg of neat MOF powder or 380 mg of MOF powder blended in an agate mortar with 20 mg of binder was introduced in the die and compressed under 3, 6, 9 or 12 metric tonnes (corresponding to pressures of 146, 292, 438 or 524 MPa, respectively) for 15 s, 30 s, 2 min, 5 min,

10 min or 15 min. For pellets with a 5-mm diameter, 190 mg of MOF powder blended in an agate mortar with 10 mg of binder was split into three equal parts and compressed under 0.3 metric tonnes (corresponding to a pressure of 146 MPa) for 15 s. The pellets were then treated at 120 °C for 2 h and exposed to air for 24 h before characterisation and testing.

3.4. Durability Tests

The durability of pellets was tested by repeatedly dropping them inside a 250-mL measuring cylinder having height of 36 cm. After each drop, the undamaged pellets were recovered by carefully turning the cylinder upside down and letting the pellet fall on a soft surface. Attrition resistance was tested by introducing about $\frac{1}{4}$ of a pellet in a 20 mL glass vial equipped with a screw cap and using a Hati Rotamixer 50 Hz vortex mixer at max power (200 W) in cycles of 30 s. After each cycle, the pellet fragment was recovered with tweezers and weighed to determine the mass loss due to attrition.

3.5. Analytical Procedures

Powder X-ray diffraction (PXRD) patterns were collected in the 4–30° 2θ range with a Bruker D8 Avance diffractometer working in reflection geometry and equipped with a LYNXEYE XE detector, using the Cu Kα radiation (Bruker, Karlsruhe, Germany). The X-ray tube was operated at 40 kV and 40 mA.

Scanning electron microscopy (SEM) micrographs of pellets were acquired with a Hitachi TM3030Plus microscope using an acceleration voltage of 5 kV (Hitachi High-Technologies, Krefeld, Germany). The samples were analysed uncoated.

N_2 sorption isotherms at 77 K were measured with a Quantachrome Nova 2000e analyzer (Quantachrome, Boynton Beach, FL, USA). The samples (about 30–50 mg) were activated for four hours under vacuum at 120 °C prior to analysis. BET surface areas were calculated in the 0.001–0.043 P/P_0 range.

3.6. CO_2 and H_2O Adsorption Tests

Dynamic CO_2 adsorption experiments and cyclic CO_2 and H_2O adsorption experiments were performed with a TA Instruments SDT-Q600 instrument (TA Instruments, New Castle, DE, USA). The samples were loaded in alumina cups and activated at 120 °C under a 100 mL min^{-1} Ar flow for 2 h. The temperature was then decreased to 40 °C and kept at this value for 30 min.

In dynamic CO_2 adsorption experiments, the gas flow was switched to CO_2 at the same flow rate and adsorption started with a little delay. In the CO_2 adsorption-desorption experiments, CO_2 was constantly flown at 100 mL min^{-1} and the temperature was swung between 40 °C and 90 °C. Each adsorption cycle lasted 40 min, whereas desorption lasted 10 min.

In H_2O adsorption-desorption experiments, a humidified air stream was achieved by bubbling dry air at 80 mL min^{-1} in H_2O at room temperature. This stream was flown during adsorption, which took place at 40 °C, whereas desorption was performed by heating to 120 °C under a 100-mL min^{-1} Ar flow. Each adsorption cycle lasted 90 min, whereas desorption took 10 min.

4. Conclusions

We have reported a systematic investigation aimed at developing an optimised compaction process for MOF-801 powders, screening parameters such as type and amount of binder, pressure and compression time. PVB was found to be the best binder among those investigated, even when blended at just 5 wt % with the MOF. The pellets compressed at 146 MPa for 15 s provided the best combination between mechanical strength and preservation of the sorption performance of the original powder over ten cycles, retaining as much as 90% of the CO_2 working capacity, while displaying unaffected sorption kinetics and 74% of the H_2O working capacity. Our work provides a starting point for future exploration of shaping of MOF powders, which is becoming a progressively more important aspect as MOFs move towards commercialisation and employment in industrial applications.

Supplementary Materials: The following are available online at http://www.mdpi.com/2304-6740/7/9/110/s1, Figure S1: PXRD pattern for pellet prepared using pre-evacuated MOF-801; Figures S2–S5: Photographs of pellets after drop tests; Figures S6 and S7: Control cyclic H_2O adsorption tests.

Author Contributions: Conceptualisation, M.T. and E.A.; methodology, M.T. and E.A.; investigation, M.T., A.G., J.L., J.S. and M.J.M.; writing—original draft preparation, M.T.; writing—review and editing, M.T., M.J.M. and E.A.; supervision, M.T. and E.A.; funding acquisition, M.T and E.A.

Funding: M.T. is supported by funding from the European Union's Horizon 2020 research and innovation program under the Marie Skłodowska-Curie grant agreement No 663830. The Engineering and Physical Sciences Research Council (EPSRC) is acknowledged for the provision of funding through the First Grant scheme (EP/R01910X/1). This work is also part of the Reducing Industrial Carbon Emissions (RICE) research operation funded by the Welsh European Funding Office (WEFO) through the Welsh Government.

Acknowledgments: The authors thank Ewa Kazimierska (Swansea University) for help with SEM analysis, Louise Hamdy (Swansea University) for help with TGA analysis and Paul M. Williams (Swansea University) for providing access to the gas sorption analyser. We would like to acknowledge the assistance provided by the Swansea University AIM Facility, which was funded in part by the EPSRC EP/M028267/1, the ERDF through the Welsh Government grant 80708, and the Sêr Solar project via the Welsh Government.

Conflicts of Interest: The authors declare no conflict of interest.

References

1. Petit, C. Present and future of MOF research in the field of adsorption and molecular separation. *Curr. Opin. Chem. Eng.* **2018**, *20*, 132–142. [CrossRef]
2. Li, H.; Wang, K.; Sun, Y.; Lollar, C.T.; Li, J.; Zhou, H.-C. Recent advances in gas storage and separation using metal–organic frameworks. *Mater. Today* **2018**, *21*, 108–121. [CrossRef]
3. Wang, Y.; Zhao, D. Beyond equilibrium: Metal–organic frameworks for molecular sieving and kinetic gas separation. *Cryst. Growth Des.* **2017**, *17*, 2291–2308. [CrossRef]
4. Valizadeh, B.; Nguyen, T.N.; Stylianou, K.C. Shape engineering of metal–organic frameworks. *Polyhedron* **2018**, *145*, 1–15. [CrossRef]
5. Nandasiri, M.I.; Jambovane, S.R.; McGrail, B.P.; Schaef, H.T.; Nune, S.K. Adsorption, separation, and catalytic properties of densified metal–organic frameworks. *Coord. Chem. Rev.* **2016**, *311*, 38–52. [CrossRef]
6. Ren, J.; Langmi, H.W.; North, B.C.; Mathe, M. Review on processing of metal–organic framework (MOF) materials towards system integration for hydrogen storage. *Int. J. Energy Res.* **2015**, *39*, 607–620. [CrossRef]
7. Bazer-Bachi, D.; Assié, L.; Lecocq, V.; Harbuzaru, B.; Falk, V. Towards industrial use of metal–organic framework: Impact of shaping on the MOF properties. *Powder Technol.* **2014**, *255*, 52–59. [CrossRef]
8. Chanut, N.; Wiersum, A.D.; Lee, U.H.; Hwang, Y.K.; Ragon, F.; Chevreau, H.; Bourrelly, S.; Kuchta, B.; Chang, J.-S.; Serre, C.; et al. Observing the effects of shaping on gas adsorption in metal–organic frameworks. *Eur. J. Inorg. Chem.* **2016**, *2016*, 4416–4423. [CrossRef]
9. Peterson, G.W.; DeCoste, J.B.; Glover, T.G.; Huang, Y.; Jasuja, H.; Walton, K.S. Effects of pelletization pressure on the physical and chemical properties of the metal–organic frameworks $Cu_3(BTC)_2$ and UiO-66. *Microporous Mesoporous Mater.* **2013**, *179*, 48–53. [CrossRef]
10. Bai, Y.; Dou, Y.; Xie, L.H.; Rutledge, W.; Li, J.R.; Zhou, H.C. Zr-based metal–organic frameworks: Design, synthesis, structure, and applications. *Chem. Soc. Rev.* **2016**, *45*, 2327–2367. [CrossRef] [PubMed]
11. Wißmann, G.; Schaate, A.; Lilienthal, S.; Bremer, I.; Schneider, A.M.; Behrens, P. Modulated synthesis of Zr-fumarate MOF. *Microporous Mesoporous Mater.* **2012**, *152*, 64–70. [CrossRef]
12. Zahn, G.; Schulze, H.A.; Lippke, J.; König, S.; Sazama, U.; Fröba, M.; Behrens, P. A water-born Zr-based porous coordination polymer: Modulated synthesis of Zr-fumarate MOF. *Microporous Mesoporous Mater.* **2015**, *203*, 186–194. [CrossRef]
13. Furukawa, H.; Gandara, F.; Zhang, Y.B.; Jiang, J.; Queen, W.L.; Hudson, M.R.; Yaghi, O.M. Water adsorption in porous metal–organic frameworks and related materials. *J. Am. Chem. Soc.* **2014**, *136*, 4369–4381. [CrossRef] [PubMed]
14. Kim, H.; Yang, S.; Rao, S.R.; Narayanan, S.; Kapustin, E.A.; Furukawa, H.; Umans, A.S.; Yaghi, O.M.; Wang, E.N. Water harvesting from air with metal–organic frameworks powered by natural sunlight. *Science* **2017**, *356*, 430. [CrossRef]
15. Ganesh, M.; Hemalatha, P.; Peng, M.M.; Cha, W.S.; Jang, H.T. Zr-fumarate MOF a novel CO_2-adsorbing material: Synthesis and characterization. *Aerosol Air. Qual. Res.* **2014**, *14*, 1605–1612. [CrossRef]

16. Wang, Y.; Li, L.; Yan, L.; Cao, L.; Dai, P.; Gu, X.; Zhao, X. Continuous synthesis for zirconium metal–organic frameworks with high quality and productivity via microdroplet flow reaction. *Chin. Chem. Lett.* **2018**, *29*, 849–853. [CrossRef]
17. Bambalaza, S.E.; Langmi, H.W.; Mokaya, R.; Musyoka, N.M.; Ren, J.; Khotseng, L.E. Compaction of a zirconium metal–organic framework (UiO-66) for high density hydrogen storage applications. *J. Mater. Chem. A* **2018**, *6*, 23569–23577. [CrossRef]
18. Graham, A.J.; Allan, D.R.; Muszkiewicz, A.; Morrison, C.A.; Moggach, S.A. The effect of high pressure on MOF-5: Guest-induced modification of pore size and content at high pressure. *Angew. Chem. Int. Ed.* **2011**, *50*, 11138–11141. [CrossRef]
19. Dhainaut, J.; Avci-Camur, C.; Troyano, J.; Legrand, A.; Canivet, J.; Imaz, I.; Maspoch, D.; Reinsch, H.; Farrusseng, D. Systematic study of the impact of MOF densification into tablets on textural and mechanical properties. *CrystEngComm* **2017**, *19*, 4211–4218. [CrossRef]
20. Terracina, A.; Todaro, M.; Mazaj, M.; Agnello, S.; Gelardi, F.M.; Buscarino, G. Unveiled the source of the structural instability of HKUST-1 powders upon mechanical compaction: Definition of a fully preserving tableting method. *J. Phys. Chem. C* **2018**, *123*, 1730–1741. [CrossRef]
21. Ren, J.; Musyoka, N.M.; Langmi, H.W.; Swartbooi, A.; North, B.C.; Mathe, M. A more efficient way to shape metal–organic framework (MOF) powder materials for hydrogen storage applications. *Int. J. Hydrogen Energy* **2015**, *40*, 4617–4622. [CrossRef]
22. Hastürk, E.; Schlüsener, C.; Quodbach, J.; Schmitz, A.; Janiak, C. Shaping of metal–organic frameworks into mechanically stable monoliths with poly(vinyl alcohol) by phase separation technique. *Microporous Mesoporous Mater.* **2019**, *280*, 277–287. [CrossRef]
23. Zheng, J.; Cui, X.; Yang, Q.; Ren, Q.; Yang, Y.; Xing, H. Shaping of ultrahigh-loading MOF pellet with a strongly anti-tearing binder for gas separation and storage. *Chem. Eng. J.* **2018**, *354*, 1075–1082. [CrossRef]

MDPI
St. Alban-Anlage 66
4052 Basel
Switzerland
Tel. +41 61 683 77 34
Fax +41 61 302 89 18
www.mdpi.com

Inorganics Editorial Office
E-mail: inorganics@mdpi.com
www.mdpi.com/journal/inorganics

www.ingramcontent.com/pod-product-compliance
Lightning Source LLC
Chambersburg PA
CBHW062039220926
51850CB00010B/2609